D0453304

pyt3 28/6/07

Perceived Exertion for Practitioners

Rating Effort With the OMNI Picture System

Robert J. Robertson, PhD

University of Pittsburgh

Human Kinetics

0 1868275

LIBRARY

ACC. No.	DEPT.
36015426	pyt3
CLASS No.	
612, C44 ROB	
UNIVERSITY OF CHESTER	
WARRINGTON CAMPUS	

28/6/07

Library of Congress Cataloging-in-Publication Data

Robertson, Robert J., 1943-
  Perceived exertion for practitioners : rating effort with the OMNI picture
system / Robert J. Robertson.
      p. ; cm.
  Includes bibliographical references and index.
  ISBN 0-7360-4837-5 (soft cover)
  1. Exercise--Measurement. 2. Exercise--Psychological aspects. 3. Perception.
4. Stress (Physiology)
  [DNLM: 1. Exertion--physiology. 2. Exercise Test. 3. Perception--
physiology. 4. Physical Fitness. 5. Rehabilitation--methods. WE 103 R651p
2004] I. Title.
  QP301.R556 2004
  612'.044--dc22

                                          2004000929

ISBN: 0-7360-4837-5

Copyright © 2004 by Robert J. Robertson

All rights reserved. Except for use in a review, the reproduction or utilization of this work in
any form or by any electronic, mechanical, or other means, now known or hereafter invented,
including xerography, photocopying, and recording, and in any information storage and
retrieval system, is forbidden without the written permission of the publisher.

Notice: Permission to reproduce the following material is granted to instructors and agencies
who have purchased *Perceived Exertion for Practitioners:* pp. 141-151. The reproduction
of other parts of this book is expressly forbidden by the above copyright notice. Persons or
agencies who have not purchased *Perceived Exertion for Practitioners* may not reproduce
any material.

Acquisitions Editor: Michael S. Bahrke, PhD; **Managing Editor:** Amanda S. Ewing;
Copyeditor: Nancy Elgin; **Proofreader:** Sarah Wiseman; **Indexer:** Betty Frizzéll; **Permis-
sion Manager:** Dalene Reeder; **Graphic Designer:** Fred Starbird; **Graphic Artist:** Denise
Lowry; **Photo Manager:** Kareema McLendon; **Cover Designer:** Keith Blomberg; **Art
Manager:** Kelly Hendren; **Illustrator:** Bill Brent; **Printer:** Versa Press

Printed in the United States of America   10 9 8 7 6 5 4 3 2 1

Human Kinetics
Web site: www.HumanKinetics.com

United States: Human Kinetics, P.O. Box 5076, Champaign, IL 61825-5076
800-747-4457
e-mail: humank@hkusa.com

Canada: Human Kinetics, 475 Devonshire Road Unit 100, Windsor, ON N8Y 2L5
800-465-7301 (in Canada only)
e-mail: orders@hkcanada.com

Europe: Human Kinetics, 107 Bradford Road, Stanningley,
Leeds LS28 6AT, United Kingdom
+44 (0) 113 255 5665
e-mail: hk@hkeurope.com

Australia: Human Kinetics, 57A Price Avenue, Lower Mitcham, South Australia 5062
08 8277 1555
e-mail: liaw@hkaustralia.com

New Zealand: Human Kinetics, Division of Sports Distributors NZ Ltd.,
P.O. Box 300 226 Albany, North Shore City, Auckland
0064 9 448 1207
e-mail: blairc@hknewz.com

This book is dedicated to those who contribute to the development of the perceived exertion knowledge base through research and teaching and to those who are committed to the application of this knowledge to promote health and fitness for individuals of all ages.

Contents

Chapter 4   Tests of Health Fitness and Sport Performance Using Rating of Perceived Exertion 33

Chapter 5   Exercise Programs Using a Target Rating of Perceived Exertion . . . 53

Preface

This book describes applications of perceived exertion as they currently are defined by professional practitioners and explores new and innovative perceived exertion programs based on the OMNI Scale of Perceived Exertion that can readily be used by movement practitioners. Chapters 1 and 2 describe the perceived exertion knowledge base and explain how the new OMNI Picture System of Perceived Exertion was developed for measuring the rating of perceived exertion (RPE). Chapters 3 and 4 explain how to use RPE in assessing physical fitness, and in particular how to identify functional and clinical performance limits. Both clinical and field-based perceptual tests to be used in assessing aerobic, anaerobic, and resistance exercise are described. Chapters 5 through 8 focus on using RPE to prescribe and regulate the intensity of exercise training, sport conditioning, and weight-loss activities. These chapters stress the practical and efficient use of *target RPE zones* and self-regulated exercise intensity by using the OMNI picture system. Finally, chapter 9, explains the use of RPE in exercise testing and prescription for individuals whose clinical or exercise status is limited by cardiac or pulmonary disease or chronic pain.

Each chapter begins with a case study that highlights a unique exercise requirement of a health-fitness client or rehabilitation patient. The case study describes the client, identifies the client's needs, and presents an action plan for meeting those needs that uses RPE as the principal exercise response. In addition, important RPE-related terms are boldfaced throughout the text. Each term is accompanied by a functional definition that is appropriate for the concepts being discussed.

The contents of this book are directed toward health-fitness, clinical, and therapeutic practitioners. However, the material is also pertinent to professionals in the fields of sports medicine, physical education, and coaching.

It is my hope that this book's introduction to the new OMNI picture system will be intellectually stimulating and of professional value to you. Please enjoy reading it and using it in your professional practice.

Acknowledgments

This book could not have been written without the invaluable assistance of Mrs. Donna Farrell, whose untiring commitment to organizational and editorial detail made publication possible. In addition, special recognition is given to Mr. William Brent for his unique and creative drawings that are used in the OMNI Scales.

1

Perceived Exertion

CASE STUDY

Client Characteristics

The client is a 35-year-old, moderately fit woman who has been participating in an aerobic exercise program for 1 year. Her 60-minute exercise class meets three times per week at a fitness club and is supervised by an instructor.

Exercise Need

During the first 6 months of her exercise program, the client experienced regular gains in fitness and looked forward to the sessions. However, the client recently reported that the training program seems to be overly structured and less pleasurable. One problem in particular is what the client says is a rigid emphasis on using a target heart rate (HR) to regulate exercise intensity. The client indicated that during exercise she often is focused exclusively on measuring her pulse and being concerned that the intensity adjustments she makes will achieve the prescribed target HR. In effect, she has become a "pulse counter," continually interrupting the flow of her exercise program to check her HR response.

Action Plan

The health-fitness practitioner told the client that an alternative method for self-regulating her exercise intensity is to use a target rating of perceived exertion (RPE) training zone. Because RPE and physiological measures such as HR and oxygen consumption provide much of the same information about exercise intensity, the client can use RPE to guide her aerobic conditioning program. The practitioner informed the client that training using target RPE zones is simple, easy to learn, and fun to use.

The perception of physical exertion involves the feelings of *effort, strain, discomfort,* and *fatigue* that a person experiences during exercise (Robertson and Noble 1997). The first scale for measuring perceived exertion was developed in the early 1960s by psychologist Gunnar Borg at the University of Stockholm (see figure 1.1). Borg's collaborators early in the development and validation of the scale included Bruce J. Noble of the University of Pittsburgh and William P. Morgan of the University of Wisconsin. Since this initial pioneering work, a number of scales for quantifying perceived physical exertion have been developed and validated. With these scales, users select the number that they feel corresponds to the intensity of their physical exertion. This number, called the **rating of perceived exertion,** or RPE, is used by exercise and clinical practitioners to describe the range of indicators that make up an individual's perception of physical exertion during aerobic or resistance exercise.

6	No exertion at all
7	Extremely light
8	
9	Very light
10	
11	Light
12	
13	Somewhat hard
14	
15	Hard (heavy)
16	
17	Very hard
18	
19	Extremely hard
20	Maximal exertion

Borg RPE scale
© Gunnar Borg, 1970,
1985, 1994, 1998

Figure 1.1 Fifteen-category Borg Perceived Exertion Scale.

Reprinted, by permission, from G. Borg, 1998, *Perceived exertion and pain scales* (Champaign, IL: Human Kinetics), 47.

PERCEIVED EXERTION KNOWLEDGE BASE

The perceived exertion knowledge base developed sequentially. First, a standardized definition of perceived exertion was developed that conformed to the accepted practices of health-fitness and clinical exercise practitioners. Next, a system for classifying and anatomically locating the origin of exertional ratings was devised. The physiological, psychological, clinical, and performance events that affect the intensity of exertional ratings within each classification were then identified and explained. Finally, the concept of perceived exertion was applied in a wide range of health-fitness, clinical, and sport settings. The scaling systems fundamental to each of these steps were developed with emphasis placed on their validity, reliability, and utility.

EFFORT CONTINUA

The functional interdependence of perceptual and physiological responses during exercise is the theoretical rationale underlying applications of RPE. This rationale is an outgrowth of Borg's original thesis that the subjective response to exercise involves three main effort continua: *physiological, perceptual,* and *performance,* as shown in figure 1.2 (Robertson 2001a). The effort continua depict the relationship between the physiological demands of exercise performance and the perception of the exertion associated with that performance. It is expected that as the intensity of exercise performance increases, *corresponding* and *interdependent* changes occur in both the perceptual and physiological processes. The functional links between the three effort continua indicate that a perceptual response provides much of the same information about exercise performance as a physiological response does. Therefore, decisions about the intensity and duration of exercise performance can be based on the functional interaction between the perceptual and physiological continua.

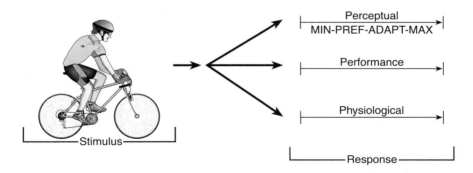

Figure 1.2   Effort continua model of perceived exertion. Min = minimum; Pref = preferred; Adapt = adaptation; Max = maximum.

Adapted, by permission, from G. Borg, 1998, *Perceived exertion and pain scales* (Champaign, IL: Human Kinetics), 6.

RANGE MODEL: CORNERSTONE OF PERCEIVED EXERTION MEASUREMENT

The use of RPE in health-fitness and clinical settings is based on Borg's range model (figure 1.3), which describes how RPE changes as exercise intensity changes from a very low to a very high level (Borg 1998). The range model makes two important measurement assumptions: that for any given exercise range between rest and maximum, there is a corresponding and equal RPE range, and that both the RPE range and the intensity of perceptions at low and high exercise levels are equal between clients regardless of their fitness status. These assumptions predict that as exercise intensity increases from minimal to maximal levels, a *corresponding* and *equal* increase from no effort to maximal effort occurs. Therefore, two clinically normal individuals can estimate their effort at 50% of the RPE range even if one of the clients has comparatively greater maximal aerobic power or muscular strength and exercised at a higher absolute intensity. RPE, then, can be compared in clients with different fitness levels.

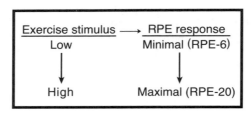

Figure 1.3 Borg's range model for category scales of perceived exertion.

Of practical importance is that the range model determines what instructions are given to clients about transposing their feelings of exertion into numerical ratings. The model also establishes procedures for setting the low and high perceptual scale anchors (see chapter 3). During these anchoring procedures, the subjective correspondence between the exercise stimulus and RPE response is established in the client's memory, setting the client's perceptual range at the same width as the exercise stimulus range. The client's perceived exertion at the low and high ends of the response continuum then corresponds directly to exercise intensities that are extremely low and extremely high.

To summarize, the range model subjectively equalizes the minimal and maximal levels of perceived exertion in individuals who otherwise vary in physiological, psychological, and physical activity attributes. The RPE at any point in the exercise stimulus range is then determined by its relative position within the response range as defined by the minimal and maximal rating scale categories.

EXERTIONAL MEDIATORS AND SYMPTOMS: AN EXPLANATION

The application of RPE in health-fitness and clinical settings requires knowledge of the underlying physiological, psychological, and symptomatic processes that clients subjectively monitor and evaluate with RPE (Noble and Robertson 1996).

Physiological Mediators

The physiological factors that influence the perceived intensity of exertion are classified as *peripheral, respiratory-metabolic,* and *nonspecific.* **Peripheral** physiological mediators are localized in the limbs and trunk. **Respiratory–metabolic** mediators are physiological responses that influence ventilatory drive during exercise. **Nonspecific** mediators are generalized or systemic physiological events that occur during exercise. Figure 1.4 lists the physiological events that research has linked to the specific

Respiratory– Metabolic	Peripheral	Nonspecific
Pulmonary ventilation	Metabolic acidosis (pH, lactic acid)	Hormonal regulation (catecholamines, β-endorphins)
Oxygen uptake	Blood glucose	Temperature regulation (core and skin)
Carbon dioxide production	Blood flow to muscle	Pain
Heart rate	Muscle fiber type	Cortisol and serotonin
Blood pressure	Free fatty acids	Cerebral blood flow and oxygen
	Muscle glycogen	

Figure 1.4   Physiological mediators of perceived exertion.
From B.J. Noble and R.J. Robertson, 1996, *Perceived exertion* (Champaign, IL: Human Kinetics).

classes. Taken together, these three classes of mediators help to determine the RPE during aerobic, anaerobic, and resistance exercise.

Physiological factors mediate perceptual signals of physical exertion by acting either individually or collectively to alter the tension-producing properties of skeletal muscle (Robertson and Noble 1997). In turn, changes in peripheral and respiratory muscle tension are monitored through a common neurophysiological pathway that transmits exertional signals from the motor to the sensory cortex. It is this neurophysiological signal that is consciously interpreted by the sensory cortex as effort sensation. The pathway for signals of perceived exertion involves a combined feed-forward and feedback mechanism (Cafarelli 1988). As exercise intensity increases, the number of central motor feed-forward commands required to increase motor unit recruitment and firing frequency in both peripheral and respiratory skeletal muscle must also increase. Corollary discharges diverge from the descending motor commands to terminate in the sensory cortex. The greater the frequency of the corollary signals, the more intense the perceived physical exertion. In addition, afferent feedback from muscles and joints helps to refine and calibrate central motor outflow commands. The resulting sensory integration of feed-forward and feedback pathways fine-tunes the exertional response.

Psychosocial Mediators

Certain psychological and sociological factors systematically influence self-assessment of effort (Morgan 2001; Noble and Robertson 1996). These factors are generally grouped into the four broad classifications listed in table 1.1.

Both experimental research and clinical observation have established that these psychosocial factors account for interindividual differences in perceived exertion. However, a consistent pattern of change in RPE in the presence of one or more of these psychosocial mediators is not evident. It is possible that the same mediator can intensify RPE in one individual and attenuate it in another. Much more research is needed in this domain of exertional perception.

Exertional Symptoms

Perception of effort is a complex psychological process that integrates a number of exertional symptoms, each of which is presumed to be linked to an underlying physiological or psychological mediator (Noble and Robertson 1996). The linguistic expression of these exertional symptoms forms a *perceptual reality* that allows for a global measurement of the physiological and psychological factors that influence exercise performance. One of the most pronounced symptoms of exertional intolerance

Table 1.1   Psychosocial Mediators of Perceived Exertion

Classification	Factor
Emotion or mood	Anxiety Depression Extroversion Neuroticism
Cognitive function	Dissociation Self-efficacy Type A personality
Perceptual process	Pain tolerance Sensory augmentation or reduction Somatic perception
Social or situational	Music Sex of counselor Social setting

From W.P. Morgan, 2001, "Utility of exertional perception with special reference to underwater exercise," *International Journal of Sport Psychology* 32(2): 137-161.

is fatigue. In addition, aches, cramps, muscular and articular pain and heaviness, and **dyspnea** (a feeling of breathlessness) are all somatic symptoms experienced during both aerobic and resistance exercise. A number of psychological symptoms—such as task aversion and low motivation—also directly affect RPE. Often, **clusters** of exertional symptoms interact to form the client's **perceptual style,** which becomes part of the **perceptual–cognitive reference filter.** This filter contains sensory information and experiences that reflect a broad range of psychosocial and cognitive processes. From an operational standpoint, the contents of the perceptual–cognitive reference filter shape the intensity of perceptual signals as they travel from their physiological/neuromotor origins to conscious expression as an RPE. In this way, the sensory content of the reference filter exerts a strong influence on the client's RPE.

PERCEIVED EXERTION TERMS AND DEFINITIONS

The following is a list of terms and definitions that are commonly used by health-fitness and clinical exercise specialists when applying RPE in their professional practice. These terms appear throughout the chapters of this book. Please refer to their definitions as necessary, especially when their meaning is important to understanding a new concept or application strategy.

> **category RPE scale**—A perceived exertion rating scale having numerical categories that represent equal intensity intervals, such as the Borg 6–20, Borg CR-10, and OMNI scales.
>
> **differentiated RPE**—An RPE for the arms, legs, and/or chest.
>
> **health fitness**—Physical fitness of the type that promotes long-term health and is derived through cardiovascular, strength, flexibility, and weight-loss activities.
>
> **perceptual signal**—A class of RPEs having either a peripheral, respiratory–metabolic, or nonspecific origin.
>
> **physiological mediator**—Physiological factors that influence the intensity of the perceptual signal (i.e., the RPE).
>
> **psycho-social mediator**—Psychological and sociological factors that influence the intensity of the perceptual signal (i.e., the RPE).
>
> **rating scale anchors**—The intensities of perceived exertion that are assigned to the lowest and highest rating-scale categories.
>
> **RPE training zone**—A stable RPE that is linked to a pre-determined physiological, psychological, or clinical training outcome.
>
> **RPE**—Rating of perceived exertion.
>
> **undifferentiated RPE**—An RPE for the overall body as a whole.

SUMMARY

This chapter describes the development of the perceived exertion knowledge base, focusing on the definition, measurement, classification, and application of RPE. Physiological and psychological mediators of RPE as well as exertional symptoms are identified and their interrelationships are explained. As the intensity of exercise performance increases, corresponding and interdependent changes occur in both perceptual and physiological responses. Therefore, RPE provides much of the same information about exercise performance that most physiological responses do. It is logical, then, that both the intensity and duration of training can be regulated using a target RPE zone. This background information provides a basis for the chapters that follow, in which the measurement and application of RPE are considered for the health-fitness client, competitive athlete, and clinical patient.

2

The OMNI Picture System of Perceived Exertion

CASE STUDY

Client Characteristics

The client is a 50-year-old woman who is moderately fit and of normal body weight. She has been a member of a health-fitness spa for 12 months. She is enthusiastic about her conditioning program and maintaining optimal levels of aerobic fitness and body weight.

Exercise Need

The client reported that she likes to use exercise stations in a combination, or combo, workout that includes a mixture of aerobic and resistance exercises performed at alternating intensities. However, she indicated that it is difficult to quickly achieve the prescribed exercise intensity when she moves to a new station in the combo circuit. Because each new exercise requires her to consciously reestablish the desired exercise intensity, her training continuity is disrupted. She requested the help of the club's fitness director in solving her exercise problem.

Action Plan

The fitness director suggested that the client use the OMNI perceived exertion scale to help her regulate her training intensity during the combination workouts. The OMNI scale is based on a single set of verbal cues in conjunction with a numerical rating range of 0 to 10. The OMNI scale also has sets of picture cues that are selected to be consistent with the exercise mode to be performed at each combo station. The client uses the pictures to help her form a mental picture of the prescribed intensity. She then easily regulates the specific intensity by exercising in the RPE zone that coincides with the appropriate picture cue. The client reported that the RPE picture system works very well, allowing her to regulate her exercise intensity with minimal disruption in her training continuity when she moves from one combo station to the next.

This book introduces the newly developed OMNI Picture System of Perceived Exertion. The term *OMNI* is short for *omnibus,* which in this context means that the perceived exertion scale is applicable for a wide range of clients and physical activity settings. The OMNI scale employs pictures of an individual exercising at different performance levels. These pictures are combined with short verbal cues and arranged along a numerical scale ranging from 0 to 10 that depicts gradually increasing exercise intensity such as that encountered when going up a hill (see figure 2.1). To be of value over a wide range of physical activities, sports, and health-fitness settings, the OMNI scale pictures depict individuals participating in different types of physical exercise. Although the pictures vary in this way, the verbal cues and their corresponding numerical ratings are always the same. The depiction to be used for a particular fitness or therapeutic session is selected by matching the exercise mode shown on the scale with the type of physical activity to be performed.

THE WHY AND HOW OF THE OMNI SCALE

The first version of the OMNI scale was constructed for children and adolescents. The Children's OMNI Scale of Perceived Exertion was developed in response to growing clinical and research interest in measuring perceived physical exertion in youths (Robertson et al. 2000a). Initially, many of these pediatric investigations employed category rating scales that had been developed for adult use. However, adult scales can pose methodological and semantic limitations when used by children and adolescents. Some children, particularly those younger than 11 years of age, cannot consistently assign numbers to words or phrases that describe exercise-related feelings. Many younger children have difficulty interpreting verbal cues that are not part of their present vocabulary. In recognition of the potential methodological and semantic limitations of such adult perceived-exertion scales as the Borg 15-category scale, the first OMNI picture system was developed for use by children and adolescents.

Following the development of the OMNI picture system for children, a version for adults was constructed. This developmental sequence was based on the expectation that lifelong participation in physical activity can be guided by the OMNI scale when picture cues are chronologically arranged to progress from childhood through adolescence and then into adulthood. It was reasoned that adults, having used the OMNI picture system when they were children, would continue to use the scale as a tool to help them select and regulate the intensity of their physical activities, which are important for both good health and recreational enjoyment. The OMNI picture system versions for children and adults have evolved into tools for guiding physical activity pursuits throughout life.

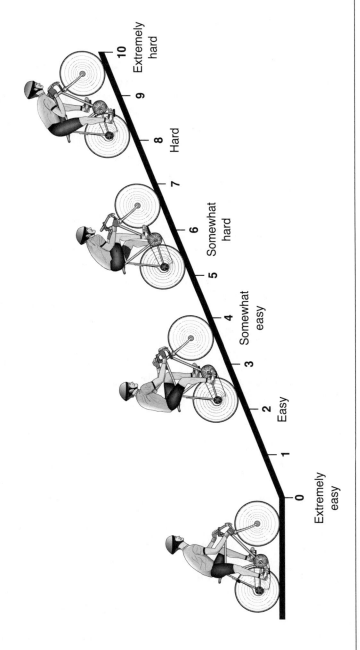

Figure 2.1 OMNI Picture System of Perceived Exertion for adult bicycle exercise.

The OMNI picture system was developed in four steps. First, an artist drew a series of pictures illustrating an individual employing various levels of exertion while performing activities such as cycling, progressing from a walk to a run, stepping, and weightlifting. A set of four pictures was drawn for each exercise, featuring both female and male children and adults. Each picture was rendered in shades of gray on a white background to maximize generalization over normal variations in human skin tones.

In the second step, several children and adults were shown the picture sets for each exercise type and asked to describe the level of physical exertion depicted by the illustrations. These verbal responses were accepted if they met one of the following criteria: (a) described effort or exertion, (b) pertained to the intensity of exercise or work, and (c) described either body signs or symptoms of exercise discomfort or comfort.

During the third step of development, semantic differential analysis was used to select from the initial pool of responses six verbal cues that each conveyed a discrete level of exertional intensity. The analysis identified one set of verbal cues that shared common meanings among children and a separate set that shared common meanings among adults. For children, the key word that was included in all of the verbal cues was *tired*. For adults, the key words were *easy* and *hard*. The verbal cues for children and adults are listed in figure 2.2.

	0	2	4	6	8	10
Adult	Extremely easy	Easy	Somewhat easy	Somewhat hard	Hard	Extremely hard
Child	Not tired at all	A little tired	Getting more tired	Tired	Really tired	Very, very tired

Figure 2.2   Verbal cues for the adult and child OMNI perceived exertion scales.

At the fourth and final step, the six semantically discrete verbal cues were placed at equal intervals along the 0 to 10 scale. The four picture cues were also positioned along the numerical rating range. This resulted in a general correspondence between the verbal and picture cues, with each depicting a discrete level of perceived exertion. The rating scale in the depiction was then raised at one end to represent a hill that required an increasing level of exertion to mount.

This four-step process was used to develop a series of OMNI scales for use by adults and children during weight-bearing and non-weight-bearing exercise. A montage of picture cues from various OMNI scale formats for both children and adults is shown in figure 2.3. A full set of OMNI scales can be found in appendix A.

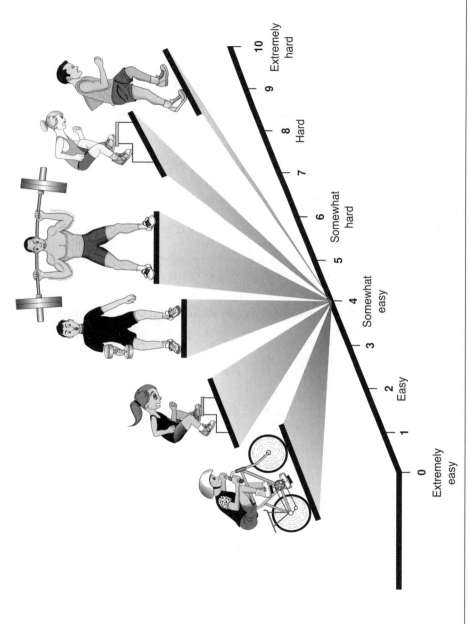

Figure 2.3 Montage of picture cues for the OMNI perceived exertion scale.

OMNI SCALE RELIABILITY

Establishing the response reliability of a newly constructed perceived exertion rating scale is a very important step in the developmental process. Pfeiffer and colleagues (2002) examined the intraclass and day-to-day reliability of both the OMNI and Borg 15-category perceived exertion scales in subjects who performed graded treadmill exercise. The reliability coefficients ranged from $r = 0.91$ to 0.95 for the OMNI scale and $r = 0.64$ to 0.78 for the Borg scale. Both scales were shown to be reliable, but the OMNI cycling format's reproducibility was superior to that of the Borg scale when used with adolescent girls.

OMNI SCALE VALIDITY

A number of research experiments have established the concurrent validity of both the child and adult versions of the OMNI scale. A concurrent validation experiment employs a criterion (that is, a stimulus) variable and a concurrent (response) variable. For aerobic exercise, for example, the most common criterion variables are oxygen consumption, HR, and power output. The criterion variables for resistance exercise are the

Table 2.1   Summary of OMNI Scale Validation for Aerobic and Resistance Exercise

		Scale	
Investigator	Mode	Format	Age
Robertson et al. (2000a)	Cycle	Cycle	Child
Utter et al. (2002)	Treadmill	Walk to run	Child
Pfeiffer et al. (2002)	Treadmill	Cycle	Child
Robertson et al. (2004)	Cycle	Cycle	Adult
			Adult
Robertson et al. (2003)	BC	Resistance	Adult
Robertson et al. (2003)	KE	Resistance	Adult

F = females; M = males; C = combined males and females; HR = heart rate; $\dot{V}O_2$ = oxygen consumption; Wt_{tot} = total weight lifted; [Hla] = lactic acid concentration; BC = biceps curl; KE = knee extension.

total weight lifted, the percentage of the one-repetition maximum (% 1RM), and the blood lactic acid concentration. The concurrent variable is always the RPE as derived from the various OMNI scales. When they were validating the OMNI picture system, researchers expected that the RPE would increase concurrently and linearly with increases in the criterion variables—oxygen consumption, HR, lactic acid, % 1RM, and so on. Because it conformed to that expectation, the OMNI scale was shown to be a valid instrument for assessing RPE in adults and children participating in both aerobic and resistance exercise. Table 2.1 summarizes this validity evidence.

INSTRUCTIONS AND PROCEDURES FOR USING THE OMNI SCALE

The OMNI scale can be used to measure RPE for the overall body (abbreviated RPE-O), the limbs (RPE-L and RPE-A), and the chest (RPE-C). To accurately rate these exertional perceptions, clients should read or listen to a recording of both the definition of perceived exertion and a standard set of OMNI scale instructions. This should be done immediately before undergoing the exercise test and again before each of the

Correlation coefficient*					
HR			$\dot{V}O_2$		
F	M	C	F	M	C
0.94	0.92	0.93	0.93	0.94	0.94
—	—	0.40	—	—	0.32
0.82	—	—	0.88	—	—
0.84	0.86	—	0.93	0.95	—
WT_{tot}		[Hla]			
F	M	C			
0.89	0.91	0.87			
0.79	0.87	—			

*All correlations, p < 0.05.

first several conditioning sessions in which a new exercise program is employed. Separate definitions of perceived exertion are used for children and adults.

> **adult**—What is the subjective intensity of effort, strain, discomfort, or fatigue that I feel during exercise?
>
> **child**—How tired do I feel during exercise?

The standard instructional set that is used with the adult OMNI scale is presented here as an example. These general instructions can be tailored to meet the specific testing and conditioning goals of your clients by changing the text to indicate the type of exercise that is to be performed. The companion instructional set for the children's OMNI scale is presented in appendix B.

Instructions

I would like you to ride on a bicycle ergometer. Please use the numbers on this scale to tell me how your body feels when you are bicycling. Please look at the person at the bottom of the hill who is just starting to ride a bicycle *(point to the left-hand picture)*. If you feel like this person looks when you are riding, the exertion will be *extremely easy.* Your rating should be the number 0. Now look at the person who is barely able to ride a bicycle to the top of the hill *(point to the right-hand picture).* If you feel like this person looks when you are riding, the exertion will be *extremely hard.* Your rating should be the number 10. If you feel like your effort falls somewhere between *extremely easy* (0) and *extremely hard* (10), give a number between 0 and 10.

I will ask you to point to the number that tells how your whole body feels, then to the number that tells how your legs feel, and then to the number that tells how your breathing feels. There are no right or wrong answers. Use both the pictures and the words to help you select a number. Use *any* of the numbers to tell how you feel when you are riding the bicycle.

After clients have read the definition and instructions, it is important to determine whether they understand how to use the scale to rate their perceived exertion. Do this by asking the following questions.

- How do you feel right now? Please point to a number on the scale.
- How do you feel when you perform your favorite recreational activity? Please point to a number on the scale.

- How did you feel when you performed the most exhausting exercise that you can remember doing? Please point to a number on the scale.

Take time to answer all of your clients' questions. This brief preparatory period will help to ensure that clients can comfortably and competently use the OMNI scale.

How to Anchor the RPE Scale

When presenting the OMNI scale to clients for the first time or reviewing it with those who need a refresher, it is helpful to establish the rating anchor points. These **anchor points** establish the perceived intensity of exertion at the low and high ends of the OMNI scale and serve as reference points to help clients use the full range of the scale's numbers in estimating their level of exertion. Being able to link the full range of scale numbers with the full range of physiological responses during exercise satisfies the requirements of Borg's range model for category rating scales, as discussed in chapter 1.

Scale anchor points should be set individually for each client. There are two primary ways to do this—the **memory procedure** and the **exercise procedure.** The memory procedure is the more practical of the two. The client is asked to think of a time when she reached a level of exertion that is equal to that depicted by the pictures at the bottom (the low anchor point) and top (the high anchor point) of the hill in the OMNI scale illustrations. During the exercise session, the client is asked to estimate her RPE by using her memory of the levels of exertion equal to the low and high anchors on the OMNI scale. For instance, if her level of exertion during exercise is about 50% of her memory of maximal physical exertion, then the RPE should fall about halfway between 0 and 10, at 5 or 6.

To use the exercise anchoring procedure, the client first undertakes a short (1 to 2 min), very low-intensity exercise bout, preferably using the same activity mode that will be performed during the test or training session. At the end of the 2-min period, remind the client that his exertion at that point should feel like that depicted in the picture at the bottom of the hill, and direct him to assign a 0 to this feeling. Next, the client should undergo a progressive exercise test ending at the point of exhaustion, or maximal intensity. Again, the same mode should be used. When the client has reached the point of maximal exercise, remind him that his exertion at that time should feel the same as that depicted in the picture at the top of the hill, and direct him to assign a 10 to this feeling. As with the memory procedure, remind the client to use the pictures and words at the low and high anchor points on the OMNI scale to guide his RPE determination. Also remind the client that if the exertion feels somewhere between the low and high anchor points, he should give a rating between 0 and 10 on the OMNI scale.

A third type of anchoring procedure involves a combination of the **memory and exercise procedures.** First, the exercise anchor points are established as a routine part of the pretraining assessment typically used to determine the client's fitness level and to establish conditioning guidelines. Once the low and high anchor points have been set during exercise, they are reinforced as needed during individual training sessions by using the memory procedure.

RATING SKILL

Rating perceived exertion is a learned skill. Therefore, as with any learning process, developing the skill to use the OMNI picture system requires time and should follow an orderly progression through three phases: (a) scale orientation, (b) practice and feedback, and (c) reinforcement.

The following is an example of how these three phases are used to familiarize a client with the OMNI scale and start the learning process to ensure that accurate ratings are given. The client is to ride a stationary cycle, so the picture cues on the OMNI scale feature a cyclist. The client is to perform three 3-min power outputs with 1 to 3 min of rest between exercise bouts. The power outputs should be selected according to the client's aerobic fitness level and progress from a low to a moderately high intensity. Before exercise, have the client review the definition of perceived exertion and the instructions for the OMNI scale. Establish the low and high scale anchor points using the memory procedure. Begin the exercise and ask the client to practice estimating her RPE (overall body and limbs) during each minute of the exercise stage. Give feedback concerning the general agreement between the RPE and the approximate exercise intensity. Regular practice with appropriate feedback will sharpen the client's rating skills.

Remember, the RPE learning process takes time. Periodic reinforcement of the OMNI scaling procedures may be helpful throughout the training program, but it is particularly important in the beginning. During the orientation session—as well as during follow-on testing and training sessions—the OMNI scale should be in full view at all times. Normally, clients are instructed to rate their perceived exertion in whole numbers only. One helpful technique is to repeat the RPE back to the client to confirm that the client thinks the rating is accurate, and, more importantly, to encourage the client to think about the veracity of the rating relative to the full exercise response range. Ask the client, "Does your rating seem to fit the exercise intensity?" Emphasize that it is all right to change the rating that was just given so that it accurately reflects the exercise intensity. This scaling drill reinforces the likelihood that the client will give appropriate RPE responses throughout the entire exercise range.

The following statements may help clients correct under- and overestimated RPE responses and reinforce appropriate RPE estimates.

- Your RPE is lower (or higher) than your heart rate tells me it should be. Please try again to estimate your RPE. Remember, the exercise intensity and therefore your heart rate will remain the same, so your RPE should be a little higher (or lower) than last time.
- Very good job. Your RPE is at about the level that we expect based on your heart rate.

LIMB AND CHEST RATINGS OF PERCEIVED EXERTION

Both children and adults can use the OMNI scale to rate the intensity of the perceived exertion in their arms, legs, and chest as well as the overall body. For most types of aerobic and resistance exercise, RPEs for the active limbs are higher than the chest RPE. The RPE-O usually falls between the RPE-C and RPE-L, making this undifferentiated rating a good global indicator of physical exertion. Differentiated RPE responses are shown in figure 2.4 as plotted OMNI scale ratings for the legs, chest, and overall body during cycle exercise. In this example, the RPE-L is more intense and the RPE-C less intense than the RPE-O. A differentiated exertional rating arising from the legs reflects peripheral perceptual signals such as acidosis, blood glucose level, and muscle blood flow, whereas a rating that is differentiated to the chest reflects respiratory–metabolic perceptual signals, including pulmonary ventilation and oxygen uptake (see chapter 1). A particularly valuable feature of the OMNI scale is the precision that it permits in distinguishing between an anatomically differentiated perceptual signal and an undifferentiated signal for the overall body when both RPEs are estimated within a comparatively narrow time

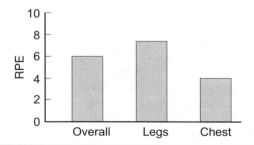

Figure 2.4 Differentiated and undifferentiated OMNI scale RPE at the anaerobic threshold.

From R.J. Robertson et al., 2001, "OMNI scale perceived exertion at ventilatory breakpoint in children: Response normalized," *Medicine and Science in Sports and Exercise* 33(11): 1946-1952.

frame. Most clients can estimate both the differentiated (limbs and chest) and undifferentiated (overall) RPE in as short a time as 30 s during either aerobic or resistance exercise.

UNIQUE FEATURES AND ADVANTAGES OF THE OMNI PICTURE SYSTEM

The OMNI scale has several distinct advantages over other perceived exertion scales that make it easier for health-fitness and clinical exercise practitioners to use. Foremost among them is that the scale employs a *single* set of verbal cues for all of the *interchangeable* sets of picture cues. The interchangeable picture cues allow the scale to be used for exercise assessment of and program prescription for clients of various ages, fitness levels, clinical statuses, and physical activity preferences. Another plus of the OMNI scale is its comparatively narrow numerical rating range of 0 to 10. Because a range of 0 to 10 is commonly used to evaluate many aspects of our daily lives, most people easily understand the scale. And finally, users report that the upper picture cue of the OMNI scale helps to sharpen their memory of maximal exertion, which often eliminates the need to engage in expensive and uncomfortable maximal exercise testing to reinforce the high scale anchor.

The measurement variability of the OMNI scale is comparatively small. The standard deviation for the RPE-O, -L, and -C is approximately ±0.6 for low, moderate, and high exercise intensities. Using a scale that has minimal measurement variability is especially important when you are prescribing an exercise program that uses a narrowly defined RPE zone.

SUMMARY

This chapter examines the development and validation of the OMNI picture system to measure RPE. The picture cues for the OMNI scale depict individuals participating in different types of physical exercise. Different sets of pictures can be selected to match the exercise to be performed, but the scale's verbal cues and their corresponding numerical ratings are always the same. The OMNI scale's flexibility makes it appropriate for use in a wide range of physical activity, sport, and health-fitness settings. The rationale underlying the need for the OMNI scale and the four-step procedure used to develop and validate the scale format are explained. The use of standardized scale instructions and anchoring procedures is emphasized, as is the importance of teaching clients how to use the OMNI scale. Finally, the measurement advantages of the OMNI scale, particularly with respect to the undifferentiated and differentiated RPEs, are explained. The OMNI scale is featured in subsequent chapters that discuss RPE applications in health-fitness, competitive, and clinical settings.

3

Traditional Methods for Rating Perceived Exertion

Client Characteristics

The client is a 38-year-old male who has been sedentary for the past 10 years and is slightly overweight, but has no other cardiovascular risk factors. The client is concerned that his low fitness level and excess body weight are affecting his cardiovascular health. He has recently enrolled in a conditioning program at his community health-fitness center.

Exercise Need

The client indicated to the fitness director that he would like help in developing a conditioning program that can be undertaken at the health-fitness center and can also be easily performed on his frequent business trips. The fitness director recommended that the client train according to target RPE zones, but the client does not have a good understanding of how to self-regulate exercise intensity using a prescribed RPE training zone.

Action Plan

The perception of physical exertion was defined for the client. He was then shown an RPE scale and asked to read a set of instructions explaining how to use the scale. The client then underwent a graded exercise test to establish the RPE anchor points and to prescribe a target RPE training zone. In addition, at the beginning of the first three to five exercise sessions at the health-fitness center, the client was reminded to use his memory of the low and high scale anchor points to achieve the target RPE training zone. This follow-on memory anchoring procedure served as a useful reinforcement tool that helped to ensure that the client would be able to self-regulate his training intensity on business trips, when he exercises away from the direct supervision of the fitness center staff.

Historically, RPE has been measured with a variety of category rating scales. The first numerical category scales were developed and validated in the early 1960s by Gunnar Borg (1998). Borg's perceived exertion and pain scales are used worldwide in the fields of exercise science, physiology, medicine, psychology, and ergonomics. Today, perceived exertion is one of the most frequently cited concepts in exercise science and sports medicine. The most widely used of the traditional RPE scales are the 1982 version of the Borg 15-category scale and the Borg Category-Ratio Scale (CR-10 scale) introduced in 1980 (Borg 1998). Borg's initial scaling systems served as measurement references for the development and validation of a number of other perceived exertion scales that are used in a wide range of exercise settings with individuals who differ in their clinical status and level of aerobic fitness. These scales include, in chronological order of development, (a) the University of Pittsburgh's 9-category scale, 1969; (b) Hogan and Fleischman's 7-category Occupational Effort Index, 1979; (c) Morgan's 7-category effort scale, 2001, 1985; (d) the Children's Effort Rating Table (CERT), 1994; and (e) the OMNI picture system, 2000 (Robertson 2001a). The scales themselves are reproduced in appendix A.

The scales were validated by correlating RPE with a corresponding HR or oxygen consumption response determined during cycle, treadmill, or water-immersion exercise tests. In addition, test and retest scale reliability was determined for short, long, and intermittent exercise protocols. The validity correlations for the scales ranged from $r = 0.56$ to 0.94. Test and retest reliability for the Borg 15-category scale, the CR-10 scale, and the Pittsburgh 9-category scale range from $r = 0.78$ to 0.91 (Robertson and Noble 1997). The reliability coefficients for the children's OMNI scale range from $r = 0.91$ to 0.95 (Pfeiffer et al. 2002). These validity and reliability coefficients indicate that numerical rating scales are acceptable instruments for assessing RPE during a wide range of aerobic and resistance exercise modes.

HOW TO SELECT A RATING OF PERCEIVED EXERTION SCALE

When measuring RPE, it is important to use scales and scaling methods that are appropriate for the individual being evaluated. As noted earlier, the assessment of exertional perceptions in health-fitness, clinical, and sport settings typically employs category rating scales. These scales require that the RPE response continuum be divided into equal intervals called *rating categories*. The scale categories are labeled with a fixed set of numbers, such as, in the Borg 15-category scale, 6 to 20. Verbal cues are then assigned to some or all of the numerical categories. The distance between each category corresponds to equal perceptual response intervals. Therefore, a change in RPE represents a change in the intensity of perceived exertion that is of equal strength from one scale category to the next.

Selecting the appropriate scale to measure exertional perceptions depends on the intended use of the RPE response. Category scales such as the OMNI picture system, the Borg 6–20 Scale, and the Borg CR-10 scale are recommended for virtually all types of exercise settings where (a) the exercise intensity is either steady or intermittent; (b) physiological responses that complement the RPE measurements change linearly with changes in exercise intensity; and (c) perceptual, physiological, or clinical responses are to be used to assess physical fitness, prescribe exercise intensity, and guide training programs. The CR-10 scale can also be used to show how perception of exertion distributes as a function of physiological responses that change exponentially with increasing exercise intensity or duration. Morgan's seven-category effort scale and the Fleishman Occupational Effort Index were developed and validated for use in, respectively, swimming and work settings.

When selecting an RPE scale, it is important to ask two questions. First, does experimental evidence document the validity and reliability of using the scale in the particular exercise setting and for the particular type of client in your program? Second, does the scale have a standardized set of instructions that explain how it is to be used? If the answer to both questions is yes, then the scale is appropriate.

HOW TO USE A RATING OF PERCEIVED EXERTION SCALE

Using an RPE scale involves (a) defining *perceived exertion,* (b) instructing the client, (c) anchoring the scale, and (d) developing the client's rating scale skills.

Perceived Exertion Definition

The definition of *perceived exertion* should present a clear and concise description of the exercise-related feelings that are to be rated. To do this, the definition should employ one or more key words. For adults, words such as *effort, strain, discomfort,* and *fatigue* form the basis of the definition. The key words are then inserted into this sentence for adults: *The perception of physical exertion is defined as the intensity of effort, strain, discomfort, or fatigue that is felt during exercise* (Robertson and Noble 1997).

Instructions to the Client

Instructions for the scale's use should be short, practical, and particular to the exercise setting. It is important that the scale instructions help the client link the full exercise stimulus range with the full RPE response range; when such a linkage is made, the basic requirements of Borg's range model have been satisfied (see chapter 1). To establish this stimulus–response linkage, the instructions should identify the lowest verbal cue, picture cue (when present), and numerical rating on the scale and state that these are equal to a very low exercise intensity. Next, the instructions should identify the highest verbal and picture cues and numerical rating and link them with maximal exercise. It is important that the instructions state whether the rating is to be undifferentiated for the overall body or anatomically differentiated to the legs, arms, and chest. The last part of the instructions should emphasize that there are no right or wrong ratings and that clients are simply to rate the intensity of their feelings at the moment. For most health-fitness and clinical settings, clients should be instructed to rate their exertion in whole numbers rather than fractions or decimals. After the client has been given the instructions, it is helpful to administer the following quiz.

- How do you feel right now? Please point to a number on the scale.
- How do you feel when you perform your favorite recreational activity? Please point to a number.
- How did you feel when you performed the most exhausting exercise that you can remember doing? Please point to a number.

The definition of *perceived exertion* and the scale instructions can be read to the client, given to the client to read, or prerecorded and played for the client. All of these options are acceptable, and the selection often depends on the client's preference. Be sure that the RPE scale you are using is in full view of the client throughout this orientation.

The following instructions are for the Borg 6–20 Scale. With slight modifications, these instructions can be used for virtually all types of category rating scales of perceived exertion.

Instructions

You are about to undergo a cycle ergometer test. The scale you see before you includes numbers from 6 to 20 and you will use it to assess your perceptions of your exertion while you exercise. The perception of physical exertion is defined as the intensity of the effort, strain, discomfort, or fatigue that you feel during exercise.

Please use this scale to translate into numbers your feelings of exertion while exercising.

The numbers on the scale represent a range of feelings from *no exertion at all* to *maximal exertion.* To help you select a number that corresponds to your exercise feelings, consider the following. When the exercise exertion feels *very, very light,* respond with a number 7. For example, you should respond with a number 7 when you are pedaling very slowly on the cycle. When the exercise exertion feels *very, very hard,* respond with a number 19. For example, a response of 19 is appropriate when your feelings of exertion are the same as when you ride a cycle almost as fast and hard as you can. If your exercise feelings are less intense than *very, very light,* respond with a number 6, and if your feelings are more intense than *very, very hard,* respond with a number 20.

You will be asked to make three separate ratings of perceived exertion during the exercise test. One rating will be for your feelings of exertion in your legs only, the second for your feelings of respiratory exertion—how hard you're breathing—in the chest, and the third for your feelings of exertion in your body as a whole. When you rate your overall exertion, be sure to select the number that most accurately represents your whole body's feelings. Likewise, when you rate the feelings in your legs or chest, pay attention only to the sensations in those specific body regions. If you feel that your effort is, for example, somewhere between 9 and 10, round your rating up to 10.

In Summary

1. You will be asked to give three ratings of perceived exertion during every minute of exercise.

2. Give each rating as accurately as possible.

3. Do not underestimate or overestimate the exertion; simply rate how the exercise makes you feel *at the moment.*

4. Use the verbal expressions to help you rate your feelings.

5. Give whatever numbers you feel are appropriate to describe your perceived level of exertion in your legs, your chest, and your total body.

SCALE ANCHORS

When clients are asked in the instructions to link the low and high ends of the RPE scale with their feelings during exercise at very low and very high intensities, they establish the rating scale's anchors. Three methods can be used to establish the anchors: the exercise procedure, memory procedure, and exercise and memory procedure.

Exercise Anchoring

To set the anchors with the exercise procedure, have the client perform a very low level of exercise. This should be the same type of exercise that will be performed during the test or training session. Instruct the client to assign the lowest numerical rating on the scale to the feelings of exertion she experiences when she performs at a very low exercise intensity. Next, ask her to perform at peak exercise intensity and instruct her to assign the highest numerical rating to her feelings of exertion at maximal intensity. After stopping the anchoring exercise, tell the client to rate her perceptions of exertion in all subsequent exercise settings relative to the feelings of the very low and maximal intensities that she just experienced.

Memory Anchoring

To use this procedure, ask the client to think about what it feels like to perform at a very low exercise intensity. Then, instruct him to assign the lowest numerical rating on the scale to the feelings of exertion that he remembers having when he performed at that low exercise level. Next, ask the client to think about how he felt when he performed at a very high exercise intensity, one so high that normally he would have to stop exercising due to exhaustion. It often helps to tell the client to remember his feelings of exertion during the most exhausting exercise that he has ever done. Instruct him to assign the highest scale rating to this memory of exertion at maximal exercise intensity.

Exercise and Memory Anchoring

To combine the exercise and memory procedures, have a new client perform exercise anchoring before she begins her conditioning program. Then, use memory anchoring to reinforce the scale anchors as it becomes necessary during the conditioning program.

SCALE ANCHORING PROCEDURES

Use the following procedures during a scale-anchoring graded exercise test (GXT) to help clients establish a firm understanding of the low and high perceptual anchor points on the Borg 6–20 RPE scale (Noble and Robertson 1996). These procedures employ an *exercise anchoring* protocol.

Low Scale Anchor

- Read to the client: "To help you identify the feelings of exertion that should be rated a 7, you will ride the cycle at an extremely low level of intensity (0 resistance) for 3 minutes."
- Explain the test protocol to the client. Begin the test, keeping the RPE scale in full view of the client.
- At 40 to 60 s into min 1 and 2, say, "Think about your feelings of exertion."
- At 40 to 60 s into min 3, say, "Think about your feelings of exertion and assign a rating of 7 to these feelings."

High Scale Anchor

- After setting the resistance level on the cycle according to your estimate of the client's fitness level, read to the client: "To help you learn the feelings of exertion that should be rated a 19, you will cycle at an extremely high level of intensity. Please exercise until you are too tired to continue."
- Explain the test protocol to the client. Begin the GXT, keeping the RPE scale in full view of the client.
- After the HR reaches 85% of the age-adjusted maximum, read the following from 40 to 60 s into each exercise minute: "When your feelings of exertion reach maximal intensity, assign a rating of 19 to these feelings." Continue until the client ends the test due to fatigue.

Immediately after the test ends, tell the client that in all future exercise tests and conditioning sessions, a rating of 6 should be given to those feelings of exertion that are less than those experienced while exercising at the extremely low intensity and a rating of 20 should be given to those feelings of exertion that are greater than those experienced during the extremely high exercise intensity. Answer any questions that the client has about the RPE scale and how it is to be used.

SCALING SKILLS: LEARNING AND PRACTICE

Using RPE to guide an exercise program is a learned skill. Learning RPE skills involves (a) orientation, (b) practice, and (c) reinforcement. Typically, the **orientation** and **practice phases** occur before or just after the client begins the conditioning program. The **reinforcement phase** is presented as needed throughout the training program. During each of these phases, the following information should be presented: (a) the definition of *perceived exertion,* (b) RPE scale instructions, (c) scale anchors, and (d) feedback.

The first three of these elements have been discussed. The fourth, feedback, to be provided to clients about their ability to use the RPE scale, is a particularly important part of the learning process. This feedback helps clients to establish a clear mental link between the intensity of the exercise and the appropriate RPE. Ideally, feedback is given in the form of *perceptual coaching* when the RPE is repeated back to the client. This encourages clients to think about the rating relative to the full exercise response range. Ask the clients, "Does your rating seem to fit the exercise intensity?" Emphasize that it is all right for them to change the rating they just gave if they decide another number would better reflect the actual exercise intensity.

IDENTIFICATION OF SCALING DIFFICULTIES

Occasionally, clients do not seem to understand the concept of perceived exertion and therefore cannot properly use a category RPE scale. How do you accurately and promptly identify these clients? An indication is often seen in the RPE responses to the preparticipation GXT. Clients who are having difficulty understanding and using an RPE scale often give ratings during a GXT similar to those shown in figure 3.1. The figure plots the RPEs (using the OMNI scale) given by three different clients during their initial GXT. All of these clients subsequently had difficulty using a target RPE to guide the intensity of their individually prescribed exercise programs. The clients' difficulty in using the category scale is evident when you compare the RPE response to a physiological response such as HR or oxygen consumption that was measured during the GXT. Client A's initial RPE was comparatively high even though the exercise intensity in the first stage of the GXT was very low. During the next few exercise stages, the client's RPE increased very rapidly to the highest scale number and remained at that rating even though the test intensity continued to increase. Obviously, this is an inappropriate response. Client B's RPE increased as the test intensity increased from low to moderate and finally to high levels. However, although Client B stopped the GXT because of exhaustion, the highest RPE was a 7 on the 10-category scale. Client C's RPE barely rose above the lowest category as the exercise intensity increased. Then, at the last test stage, the client gave an RPE of 10 and abruptly terminated the exercise test.

It is important to distinguish between clients who are having rating difficulties and those whose RPE responses exhibit normal fluctuations either within a given training session or from one session to the next. One way to distinguish between these two clients is to examine their lowest and highest RPE responses. These responses can be obtained during an exercise test or a daily conditioning session. Clients A, B, and C obviously

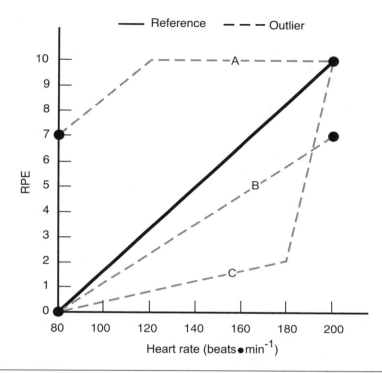

Figure 3.1 Ratings of perceived exertion given by three clients (A, B, and C) who were having difficulty using the RPE scale. The solid is the expected (i.e., reference) RPE response.

used the terminal (lowest and highest) scale numbers inappropriately, indicating that they had not been able to link the full exercise intensity (stimulus) range with the full RPE (response) range. In the case of each of these clients, then, the basic requirement of Borg's range model for RPE scaling was not met (see chapter 1).

SOLUTIONS FOR THE HEALTH-FITNESS CLIENT

Clients who are having difficulty with using the RPE scale can be taught to use the scale correctly by undertaking the three-point approach to learning scaling skills: orientation, practice, and reinforcement. Clients typically have the most difficulty with the RPE scale when they attempt to regulate their exercise intensity according to a target RPE. For these individuals, both a target RPE and a target HR should be calculated. During the first few minutes of the training session, have clients focus on regulating their exercise intensity by using the target HR. Simultaneously, coach them to link the appropriate target RPE with the feelings of exertion they are experiencing at that moment. Over the course of the training session, place less emphasis on the target HR and instead use the target RPE as the primary guide for regulating the training intensity.

SOLUTIONS FOR THE RESEARCH SUBJECT

The procedures just outlined are appropriate for use in health-fitness and clinical settings, but not research settings. In research experiments, it is essential that subjects who have scaling difficulties be identified and removed from the study before beginning data collection. Identifying and eliminating these subjects from the study at the conclusion of the investigation pose statistical and ethical problems. Means for identifying subjects with scaling difficulties should be concisely stated in the subject inclusion and exclusion criteria for the investigation and stipulate that subject candidates must demonstrate the cognitive ability to link the full exercise-stimulus range with the full perceived exertion response range. All subjects in the study will then meet the basic requirements of Borg's range model.

In research studies, some individuals may have cognitive or neurosensory deficits that prohibit them from using an RPE category scale according to normal procedures. The perceptual responses of this interesting class of subjects may yield significant information about neurosensory deficits that is of use to both exercise clinicians and health-fitness practitioners. However, these individuals should not be entered into subject groups with individuals whose perceptual responses are consonant with the predictions of Borg's range model. Instead, they should be studied as a special class of research subjects.

Borg said that "man reacts to the world as he perceives it not as it really is" (1998). In short, when a scale is used correctly, there are no right or wrong RPEs. The rating simply reflects how the client felt at the moment the RPE was made. Nothing else is required for clients to use target RPE zones to guide their health-fitness programs.

IS THERE A DOMINANT RATING OF PERCEIVED EXERTION?

RPE can be individually measured for the limbs, shoulders, and chest as **differentiated RPEs.** In contrast, an **undifferentiated RPE** measures feelings for the overall body. Both differentiated and undifferentiated RPEs can be assessed in a relatively short time (about 30 s) for most types of aerobic and resistance exercise. When a number of differentiated RPEs are assessed during the same exercise session, one of these ratings usually is the most intense and, therefore, the *dominant RPE* for that period. The dominant RPE often can be used to select the exercise intensity that is perceptually preferable to the client and also does not exceed the client's functional and clinical exercise tolerance. For example, selecting a target RPE for the legs may be appropriate for the client who performs cycling activities, whereas a target RPE for the arms may be appropriate when the client performs an upper body resistance exercise set. In this way, using the dominant RPE to prescribe the exercise intensity promotes muscle-specific training gains and minimizes the risk of exercise-induced injuries.

In addition, because the limb and chest ratings provide more precise information about anatomically regionalized perceived exertion, they are particularly useful in guiding exercise rehabilitation for neuromuscular, articular, and pain disorders. RPEs differentiated to specific body regions such as the legs and chest, for example, are often used in limb-specific exercise prescriptions and in the clinical assessment of exertional dyspnea, respectively.

The RPE that is the dominant response is determined by (a) the exercise type, (b) the anatomical origin of the differentiated feelings (the arms, legs, shoulders, or chest), and (c) the performance environment (air or water at a hot, cold, or neutral temperature). For example, during stationary cycling the differentiated RPE-L provides the dominant perceptual cue, whereas during arm exercise, the RPE-A is dominant (Robertson and Noble 1997). For treadmill walking or running, the RPE-L is dominant; for hand-weighted bench stepping and arm and leg water exercise, it is the RPE-A; and for carrying a weight (such as a box or bag) while walking, the dominant RPEs are those for the arms and chest. Therefore, the dominant RPE should be identified separately for physical activities that vary in type, intensity, and the specific body regions involved in the exercise.

Interestingly, the RPE-O is often very close to a mathematical average of the various differentiated RPEs. This suggests that the separate perceptual cues arising from the exercising body regions combine to form the undifferentiated RPE. RPE-O, then, is a good general indicator of the separate exertional feelings experienced in the specific body regions that are active during exercise. In fact, during some types of exercise in which both the upper and lower portions of the body are involved, the RPEs for the arms and legs are equal. When this is the case, however, the undifferentiated RPE is actually greater than the differentiated RPE-A and RPE-L. This effect has been shown for exercises performed using such combined arm and leg machines as climbers and rowers and for selected work tasks like pushing a wheelbarrow (Borg 1998).

It is important methodologically in using category rating scales of perceived exertion that the highest differentiated and undifferentiated rating category be reached at or at about the same time that maximal exercise intensity is reached. The highest rating categories (19 or 20 on the Borg scale) must correspond to the maximal exercise intensity to agree with Borg's range model and to allow RPE comparisons to be made between exercise sessions for the same client and between different clients. The range model predicts that the overall rating will be in the highest scale category, as will the limb and chest ratings when exertion is maximal. When RPE responses match these predictions, it is evident that the scale anchoring procedures were understood and correctly applied by the client.

SUMMARY

In this chapter, traditional procedures used to assess RPE with category scales such as the Borg 6–20 and CR-10 scales are examined. Preparing a client to use an RPE scale involves (a) defining *perceived exertion,* (b) instructing the client, (c) anchoring the scale, and (d) developing the client's rating scale skills. Remember that these rating scale procedures are learned best when they are combined with practice and reinforcement. Tips are provided for identifying and helping clients who are having trouble learning to estimate RPE. Finally, the conditions under which it may be advisable to use an RPE that is differentiated to the active muscles to guide training intensity are described.

4

Tests of Health Fitness and Sport Performance Using Rating of Perceived Exertion

CASE STUDY

Client Characteristics

The client is a 28-year-old, clinically normal woman who has been participating in a cardiovascular conditioning program at a health-fitness club for 2 months. She did not participate in structured exercise before joining the club. She enjoys exercising and has made a commitment to attend two exercise sessions during the week and one session on both Saturday and Sunday.

Exercise Need

Because the client is adhering strictly to the prescribed exercise program, she is experiencing rapid gains in her cardiovascular fitness. To determine that the client's prescribed exercise dosage is appropriate, the club staff must make frequent assessments of her training progress without disrupting the client's normal daily training.

Action Plan

The exercise practitioner will administer a single-level cycle test every 2 weeks to determine the client's R_{PO} score—the RPE measured during the last minute of a single 5-min power output (PO) on a cycle ergometer. As the client's cardiovascular fitness improves, her R_{PO} score will decrease if she is progressing satisfactorily and the prescribed exercise dosage is appropriate for her training status. Being informed about her training progress will help the client stay enthusiastic about her fitness program and committed to regular training sessions. In addition, frequent testing will provide the practitioner with information about the accuracy of the client's exercise prescription and whether it requires adjustment.

Exercise tests that employ RPE are used in health-fitness, sport, and rehabilitation settings for (a) measuring aerobic and anaerobic fitness, (b) prescribing individual exercise programs, and (c) tracking training progress. The following sections examine procedures that use RPE for assessing physical fitness, prescribing exercise training programs, and regulating exercise intensity for clinically normal clients and rehabilitation patients. The RPE-based tests that are described have both submaximal and maximal end points and are applicable for aerobic, anaerobic, and resistance exercise.

Combining objective physiological measures with subjective psychological measures during a GXT provides robust measurements of the total body strain imposed by exercise and the capacity of the client to tolerate that strain. As noted in chapter 1, a number of interrelated physiological factors influence RPE. Therefore, RPE responses to a GXT provide much of the same information as physiological test responses do. Measuring RPE during exercise testing is inexpensive and noninvasive and requires no bioelectrical instrumentation. A single RPE scale can be used to test clients of differing ages, genders, athletic types, and clinical statuses. In addition, many individuals can be tested in a short period. This is especially true when RPE is measured during tests that have a submaximal end point, such as those that are periodically administered during health-fitness and therapeutic training programs.

RATING OF PERCEIVED EXERTION AS A TEST GUIDE

One of the most common and oldest applications of RPE in the health-fitness setting is its use in guiding the course of a GXT to the test's end point. In this situation, RPE measurement serves two very important functions. First, the RPE at any given test stage indicates the point within the client's exercise range (rest to maximum) to which the evaluation has progressed. When the RPE reaches the upper third of the rating scale, the initial procedures to safely terminate the test should be put into place. This warning of the test's impending termination ensures that the practitioner will have adequate time to measure the physiological and clinical responses that indicate that the upper limits of the client's exercise capacity have been reached. Accurate measurements at the test's end point are necessary to develop an effective exercise program and to assess training progress. Second, when the client gives the highest RPE on the scale, the practitioner should take it as subjective confirmation that the physiological or clinical end points of the GXT have been reached.

Figures 4.1 and 4.2 show the RPE *warning zones* that indicate impending test termination. When the RPE is 15 to 17 on the Borg scale or 7 or 8 on the OMNI scale, preliminary procedures to safely end the GXT should be initiated (Noble and Robertson 1996). Monitoring the RPE for the increase that is expected to occur with increasing increments of exercise, listed in table 4.1, will help practitioners estimate their clients' next RPE as the intensity of the GXT progresses from one stage to the next (Noble and Robertson 1996). This information can be used to project when the RPE will fall within the zone that warns of test termination.

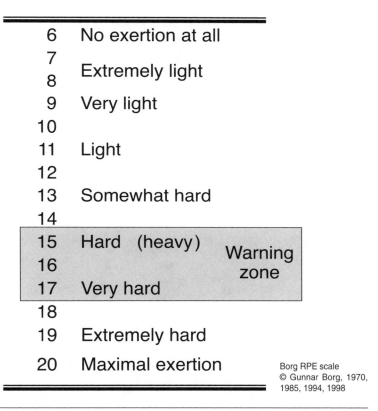

6	No exertion at all
7	
8	Extremely light
9	Very light
10	
11	Light
12	
13	Somewhat hard
14	
15	Hard (heavy)
16	Warning zone
17	Very hard
18	
19	Extremely hard
20	Maximal exertion

Borg RPE scale
© Gunnar Borg, 1970, 1985, 1994, 1998

Figure 4.1   RPE warning zone that signals impending exercise test termination, Borg 6–20 Scale.

Reprinted, by permission, from G. Borg, 1998, *Perceived exertion and pain scales* (Champaign, IL: Human Kinetics), 50.

Guiding test progression with RPE is especially helpful when assessing the symptom-limited functional aerobic power of cardiac and pulmonary patients who are taking medications that can alter HR. For example, in patients taking the cardiac medications propranolol or atropine, these drugs can respectively reduce or accelerate the HR response to aerobic

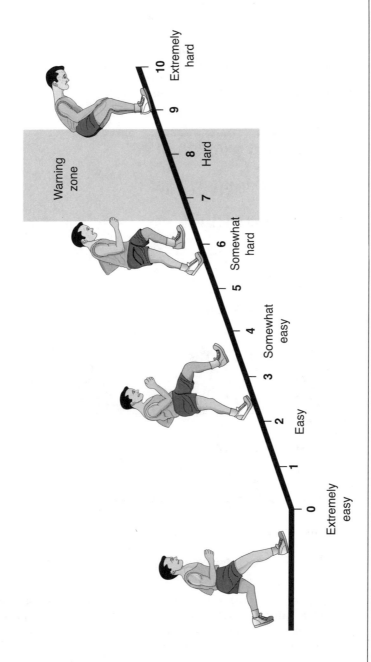

Figure 4.2 RPE warning zone that signals impending exercise test termination, OMNI scale.

Table 4.1   Expected Increase in RPE During Graded Exercise Testing

Mode	Intensity increment	RPE increment	
		Borg 6–20 Scale	OMNI scale
Treadmill			
Walking	1.0 mile/hr^{-1}	1–2	1
Running	1.0 mile/hr^{-1}	1	1
Cycle ergometer	50 W	1	1

exercise. When this occurs, the measurement of HR may not be useful in signaling the test end point. However, because RPE is largely unaffected by cardiac medications, it is an effective tool for monitoring test progression and establishing test termination criteria for coronary patients.

ASSESSING AEROBIC FITNESS

RPE can be used as the principal measurement method during both laboratory and field-based tests to estimate aerobic fitness and track conditioning progress.

Submaximal RPE Cycle Tests

RPE responses to multilevel and single-level submaximal cycle ergometer tests can be used to estimate clients' aerobic fitness or to track changes in fitness as training progresses (Noble and Robertson 1996). In the **multilevel** cycle test, the client's RPE at various levels of PO (measured in watts) are identified by using a progressively incremented exercise protocol. The higher the PO at the criterion RPE, the higher the client's level of aerobic fitness. In the **single-level** cycle test, the client reports the RPE at a criterion PO. A lower RPE at the criterion PO indicates a higher aerobic fitness level. Both of these cycle tests are useful in tracking changes in aerobic fitness as clients adapt to their conditioning programs.

Multilevel Cycle Test

Aerobic fitness can be estimated using RPE responses during a multilevel submaximal cycle ergometer test. RPE is measured during the final 15 s at each exercise level or stage. Follow these steps when administering the test.

- Set the initial PO at 25 W for females and 50 W for males. The RPE varies with the pedal rate at a constant PO. Therefore, the pedal rate should be 50 revolutions per minute (rpm) throughout the test regardless of whether an electronic or mechanically braked ergometer is used.
- Increase the PO by 25 W for females and 50 W for males at the beginning of each 3-min exercise level or stage.
- Use of the multilevel protocol set out in table 4.2 is recommended for the RPE cycle test.
- Ask the client to give an RPE (using either the Borg or OMNI scale) during the last 15 s of exercise at each level or stage. (However, you may want to measure RPE at the end of *each* exercise minute to allow clients to practice their rating skills).
- The test continues until the client gives an RPE of 17 on the Borg scale or 8 on the OMNI scale.

Table 4.2   Multilevel Protocol for the RPE Cycle Test

		Test PO level (W)*		Record	
Stage	Time (min)	Female	Male	RPE	HR
I	0–3	25	50		
II	3–6	50	100		
III	6–9	75	150		
IV	Continue PO progression until test termination.				

*Pedal rate: 50 rpm

The test results are used to calculate the client's aerobic fitness.

- The RPE measured during the last minute of exercise at each level or stage is plotted against its corresponding PO value as shown in figure 4.3.
- A reference line that represents the best fit is then drawn through the data points.
- A horizontal line is drawn from a criterion RPE (usually 15 on the Borg scale and 7 on the OMNI scale) on the y-axis to the reference line and a vertical line is drawn from the point of intersection on the reference line to the x-axis.

- The point where the vertical line meets the x-axis identifies the PO the client had reached at that criterion RPE.
- When the criterion RPE is 15 on the Borg scale, this value is called the PO at RPE of 15 (PO_{R15}); when it is 7 on the OMNI scale, it is the PO at RPE of 7 (PO_{R7}).

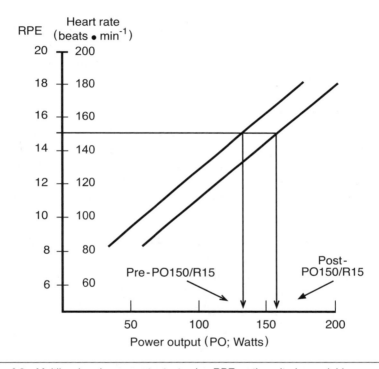

Figure 4.3   Multilevel cycle ergometer test using RPE as the criterion variable.

Meaning of PO_R

The multilevel RPE cycle test can be administered to the client immediately before beginning and after completing an aerobic conditioning program. If the PO_R increases from the pre- to post-training test, it indicates that the client's aerobic fitness level has also increased. This test was validated at the Center for Exercise and Health-Fitness Research (CEHFR) at the University of Pittsburgh. Validation correlations between PO_{R15} and peak oxygen consumption on a cycle are high and statistically significant ($r = 0.92$).

The multilevel RPE cycle test has a number of advantages that make it ideal for use in the health club and clinic. Most importantly, its use of a submaximal end point means that the test is low risk and does not require medical supervision. In addition, it is short, does not require complex physiological monitoring, and needs only minimal testing equipment (just a cycle ergometer and an RPE scale). Therefore, the test can be administered frequently and economically throughout the course of the client's conditioning program. Frequent testing provides the practitioner with information about the appropriateness of the client's exercise prescription and whether adjustment is required. In addition, frequent assessments give the client feedback about training progress and aerobic fitness gains, motivating the client to stay with the exercise program. This is especially true when clients can frequently feel and see the health benefits of regular participation.

Single-Level Cycle Test

The single-level cycle test measures RPE at a designated submaximal PO. This simple assessment can be used to track clients' training progress, alerting the practitioner to adjust the exercise dosage to maintain an overload training stimulus. The single-level cycle ergometer test is administered at the POs listed in table 4.3 and using the following procedures.

- Select the test PO level—low or high fitness—based on your estimate of the client's aerobic fitness.
- Have the client perform at the warm-up PO for 2 min followed immediately by 5 min at the test PO level.
- Ask the client for an RPE during the final 45 to 60 s of each minute of exercise at the test level. The RPE during the fifth minute is called an R_{PO} score.
- Instruct the client to cool down for 2 min at the PO indicated in table 4.3.

Table 4.3   Power Output for a Single-Level Cycle Test

	Test PO level (W)			
	Warm-up (2 min)	Low fitness (5 min)	High fitness (5 min)	Cool-down (2 min)
Female	25	50	100	25
Male	50	100	150	25

Interpreting the Single-Level Cycle Test Score

The RPE at the end of the 5-min test is the client's R_{PO} score. For example, if the client gives an RPE of 5 when using the OMNI scale (or of 12 when using the Borg scale) at a PO of 100 W, the test score is $R_{100} = 5$. As aerobic fitness improves with training, the R_{PO} score decreases. As a general rule, a decrease in the R_{PO} score of 1 OMNI scale unit or 2 Borg scale units indicates an improvement of approximately 15% in aerobic fitness. As noted previously, when the R_{PO} score improves, it is time to increase the client's training dosage. Remember that the same test PO level is used throughout the client's training program, so a lower R_{PO} score indicates that the client is able to perform the same level of submaximal work with less effort—a sign of improved aerobic fitness.

RPE Run Test

The RPE run test is a submaximal evaluation that can be easily administered as part of a daily training program to assess clients' progress and classify their aerobic fitness (Borg 1998). Periodic assessment of aerobic fitness is an important part of an exercise training program. Unfortunately, many aerobic fitness tests require expensive, time-consuming, and electronically complex procedures that allow practitioners to evaluate only one client at a time. These involved testing procedures may not be practical in some health-fitness, clinical, and physical education settings in which assessments must be undertaken on a gymnasium floor or running track. The RPE run test was designed to overcome many of these practical limitations.

Test Procedures

Instruct the client to perform at three different, self-selected jogging or running speeds in a single assessment session.

- Trial 1: Instruct the client to run very slowly at a constant speed.
- Trial 2: Have the client run a little faster, at a moderate but comfortable pace.
- Trial 3: Ask the client to run a little faster than during trial 2. Expect the running speeds selected for each trial to differ between clients. In general, clients with better aerobic fitness select faster running speeds.
- The distance for each trial can range from 220 to 660 yd. Choose the distance to make the trial last from 2 to 5 min. In general, the more sedentary the client, the shorter should be the test distance.
- Using a measured running track or gymnasium floor for the test

helps to ensure that the required test distance is traversed during each trial.

- Position a large RPE scale at the finish line. As clients cross the line, ask them to rate their RPE-O.
- Record the RPE and the amount of time it took the client to complete the trial.
- Allow a 1- to 5-min recovery period between trials. Then begin the next trial, reminding the client to run faster than during the previous trial.

Calculating the Test Results

- The first step in calculating the test responses is to convert the time taken to run the trial distance to miles per hour (mph). For a distance of

$$110 \text{ yd, } mph^{-1} = 3600 \div (\text{lap time in seconds} \times 16);$$
$$220 \text{ yd, } mph^{-1} = 3600 \div (\text{lap time in seconds} \times 8); \text{ and}$$
$$440 \text{ yd, } mph^{-1} = 3600 \div (\text{lap time in seconds} \times 4).$$

- Next, plot the RPE against its corresponding running speed for each trial by drawing a reference line to connect the RPE data points, as shown in figure 4.4.

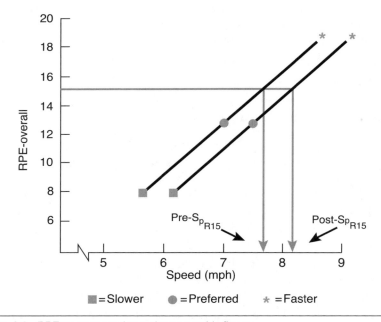

Figure 4.4   RPE run test used to estimate aerobic fitness.

- Select a target RPE, usually 15 on the Borg 6–20 Scale (Borg 1998). On the plot, draw a horizontal line from the target RPE to the reference line.
- Draw a vertical line downward from the point of intersection on the reference line to the x-axis. The point where the vertical line meets this axis is the client's running speed at the target RPE of 15 on the Borg scale (Borg 1998). This speed can be abbreviated as Sp_R. The faster the Sp_R, the better the client's level of aerobic fitness.

The run test can be administered before and immediately after the client has participated in an aerobic training program. If the client's Sp_R has increased in the post-training test, so too has the client's estimated aerobic fitness level. Borg validated this test, showing that Sp_R correlates strongly with measures of physical work capacity, $r = 0.74$ (Borg 1998). In addition, validation experiments at CEHFR have shown that the Sp_R correlates strongly with treadmill maximal oxygen consumption, $r = 0.92$.

Run Test Nomogram

The nomogram in figure 4.5 is an extension of the RPE run test just described and can be used to predict the client's maximal oxygen consumption, or $\dot{V}O_2max$ (Robertson et al. 1994). The nomogram simplifies the run test by using the RPE response to only one running speed of the client's choosing according to his preference and level of condition. To build the nomogram,

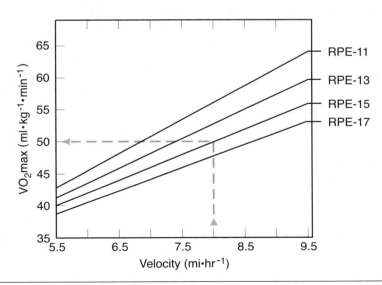

Figure 4.5 Nomogram to predict $\dot{V}O_2max$ from running velocity at a criterion RPE.

- measure the RPE using the Borg 6–20 Scale and calculate the running speed, as described above, for only one self-selected trial;
- find the running speed on the horizontal axis of the nomogram and extend a vertical line upward until it intersects with the diagonal line that plots the client's RPE for that speed; and
- extend a horizontal line from the point of intersection on the diagonal line to the y-axis.

The point where this line crosses the vertical axis indicates the client's predicted $\dot{V}O_2$max.

RPE Walk Test for Clinical Assessment

Six-minute walking tests are often administered in hospital settings to assess the functional status of coronary patients (Eng et al. 2002). Alternatively, instead of a fixed time limit, the assessment can be terminated when the patient's RPE reaches 12 on the Borg scale (Bar-Or 1977). The length of time it takes the patient to reach that RPE is taken as a measure of the patient's functional exercise tolerance. This type of test allows patients to select a pace that they feel they can tolerate for approximately 6 min. However, because the test ends at a predetermined RPE and not after a fixed time limit, test administration is simpler and patient safety and comfort are easier to maintain.

WORKPEAK TEST OF ANAEROBIC POWER

The *workpeak* test was developed by Borg (1998) to estimate anaerobic power using submaximal RPE responses to a cycle ergometer protocol. The workpeak test predicts the highest PO that a client can maintain on a cycle ergometer for 30 s—that is, the peak anaerobic power. Administer the workpeak test as follows.

- Use a cycle ergometer that has a speedometer that displays revolutions per minute and accurately adjusts PO expressed as watts.
- Have male clients complete the initial 30 s PO at 50 W and female clients, at 25 W. Instruct the client to pedal at 50 to 60 rpm and to watch the speedometer to monitor the pedal rate.
- Increase the PO by 50 W for men and 25 W for women for each subsequent 30 s period.
- Ask the client for her RPE at the end of each 30 s stage.
- Stop the test when the client's RPE reaches 17 on the Borg 6–20 Scale.

How to Calculate Workpeak

The RPE responses corresponding to each PO level are used to predict the workpeak, the client's peak anaerobic power. Figure 4.6 shows a peak power prediction using an RPE measured with the Borg 6–20 Scale.

- First, plot each RPE against its corresponding PO.
- Connect the points on the plot by approximating a straight line from the lowest to the highest RPE.
- Extend the straight line until it intersects a horizontal line drawn from RPE 20.
- Finally, draw a vertical line from this point of intersection until it crosses the x-axis, which indicates the workpeak value.

Borg reported strong validity correlations ($r = 0.96$) for the workpeak test (Borg 1998).

LIBRARY, UNIVERSITY OF CHESTER

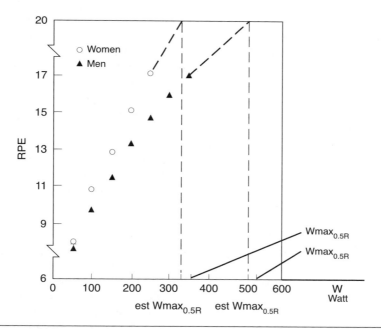

Figure 4.6 Cycle ergometer workpeak test using RPE as a criterion variable.

Reprinted, by permission, from G. Borg, 1982, "Rating of perceived exertion and heart rates during short-term cycle exercise and their use in a cycling strength test," *International Journal of Sports Medicine* 3: 153-158.

Predicting Test Performance

In both health-fitness and sport-training settings, it is sometimes practical to use a prediction technique (model) to estimate the client's or athlete's capacity for exercise performance (Noble and Robertson 1996). This can be done rather easily with equations that use RPE responses to short-duration, low-intensity tests. Prediction models based on RPE can be used to estimate the client's oxygen consumption ($\dot{V}O_2$), HR, peak PO, and endurance time. The tests that employ these prediction models typically are easy to administer and require very little time to complete. Therefore, they can be used frequently to judge training progress, adjust exercise dosage, and provide progress reports to the client. See table 4.4 for examples of equations that are used to predict exercise responses.

RPE Measurement During Resistance Exercise

Measuring RPE responses to resistance exercise has become an important tool in health-fitness, sport, and clinical rehabilitation settings. This RPE application is comparatively new, with its popularity paralleling the ever-increasing number of clients who undertake strength training in recreational fitness and work-hardening programs (Robertson et al. 2003). In these exercise settings, beginning exercisers need a simple-to-use, subjective guide to help them determine physiologically appropriate and safe resistance training intensities. RPE meets this need.

RPE assessed during resistance exercise can be used to (a) prescribe training intensity, (b) guide day-to-day training dosage, and (c) track training progress. Each of these resistance exercise applications depends on an accurate assessment of RPE for the specific upper and lower body muscles to be trained.

Let's consider an example of RPE measurements that distinguish between different resistance exercise intensities. A common way to express resistance exercise intensity is as a percentage of the one-repetition maximum (% 1RM). Lagally et al. (2002) measured RPE in young, recreationally active weightlifters who performed a series of resistance exercises at 30 and 90% 1RM and found that the RPEs for the active muscles and the overall body were higher during the 90% than during the 30% 1RM sets. This reveals two important points. First, when appropriate scaling instructions and testing procedures are used, it is possible to simultaneously measure RPE for the active muscles and the overall body during a single exercise repetition or following a set of repetitions. Second, during resistance exercise, RPE accurately distinguishes between different intensities, or % 1RM. Therefore, RPE can be used in tests of muscular strength or endurance and in the subsequent development of perceptually based resistance exercise training programs (see chapter 6).

Testing With the OMNI Resistance Exercise Scale

A special feature of this book is the introduction of the OMNI picture system to measure perceived exertion. The development and validation

Table 4.4 Exercise Prediction Equations That Use RPE

Exercise prediction/mode/units of measure	Gender/age	Equation	Scale	Correlation	Source
$\dot{V}O_2$max/treadmill/ml · kg^{-1} · min^{-1}	M and F/adult	-3.94(RPE-O) + 96.2	Borg 6–20	$r = -0.81$	Robertson et al. 1994
Arm endurance/min	M/adult	-30(RPE-A) + 303.2	Borg CR-10	$r = -0.79$	Robertson et al. 1990b
$\dot{V}O_2$/cycle/l · min^{-1}	F/adult	0.14(RPE-O) + 0.77	OMNI	$r = 0.88$	Robertson et al. 2003
$\dot{V}O_2$/cycle/l · min^{-1}	M/adult	0.30(RPE-O) + 0.33	OMNI	$r = 0.94$	Robertson et al. 2003
$\dot{V}O_2$/cycle/ml · min^{-1}	M and F/ child	124.58(RPE-O) + 166.78	OMNI	$r = 0.94$	Robertson et al. 2000a
HR/cycle/beats · min^{-1}	F/adult	7.36(RPE-O) + 107.59	OMNI	$r = 0.83$	Robertson et al. 2003
HR/cycle/beats · min^{-1}	M/adult	10.83(RPE-O) + 67.33	OMNI	$r = 0.90$	Robertson et al. 2003
HR/cycle/beats · min^{-1}	M and F/child	9.90(RPE-O) + 81.09	OMNI	$r = 0.93$	Robertson et al. 2000a
Wt$_{tot}$/BC/kg	F/adult	0.05(RPE-AM) + 1.45	OMNI	$r = 0.89$	Robertson et al. 2003
Wt$_{tot}$/KE/kg	F/adult	0.01(RPE-AM) + 4.76	OMNI	$r = 0.79$	Robertson et al. 2003
Wt$_{tot}$/BC/kg	M/adult	0.02(RPE-AM) + 1.88	OMNI	$r = 0.91$	Robertson et al. 2003
Wt$_{tot}$/KE/kg	M/adult	0.01(RPE-AM) + 3.58	OMNI	$r = 0.87$	Robertson et al. 2003

$\dot{V}O_2$max = maximal oxygen consumption per minute; M = male; F = female; RPE-O = overall; RPE-A = arms; $\dot{V}O_2$ = oxygen consumption; HR = heart rate; Wt$_{tot}$ = total weight lifted for one set (10 repetitions) of biceps curls (BC) and knee extensions (KE); RPE-AM = active muscles.

of the OMNI scale is discussed in chapter 2. One set of OMNI scale pictures depicts a weightlifter (figure 4.7), and it is intended for use during all types of resistance exercise (Robertson et al. 2003). The male and female children's versions of this set appear in appendix A, and a specially modified set of instructions for their administration is presented in appendix B.

Both men and women can use the OMNI Resistance Exercise Scale (OMNI-RES) to measure their RPE during upper and lower body resistance exercise. For example, in a study at the University of Pittsburgh's CEHFR, a group of female and male recreational weightlifters performed biceps curls and knee extensions in sets of 4, 8, and 12 repetitions and measured their RPE with the OMNI-RES. The total weight that was lifted increased from one set to the next, as did the RPEs for the active muscles (RPE-AM) and the overall body. For each set of upper and lower body exercise, the RPE-AM was always higher than the RPE-O. The responses were similar in female and male weightlifters, validating the health-fitness applicability of the OMNI-RES.

Of practical importance is that the subjects were able to use the OMNI-RES to separately estimate their RPE-AM and RPE-O in assessments that took place within a 3 s period. This study demonstrated that comparatively rapid assessment of these two RPEs is possible during multiple-set protocols that use different muscle groups. Being able to refer to an RPE that is anatomically regionalized to the primary muscles required for performance increases the precision of a perceptually guided resistance exercise program.

The following is an example of how RPE-AM and RPE-O can be assessed during upper and lower body resistance exercise (Gearhart et al. 2001).

- Select the resistance exercise that is to be used in the assessment and explain the weightlifting technique to the client.
- Determine the number of repetitions to be done for each exercise set.
- Instruct the client to assess mentally the RPE-AM during the concentric phase of the muscle action. The client will then respond with the rating during the eccentric phase.
- RPE-AM typically is measured during the middle and final repetitions of an exercise set.
- During the final repetition and immediately after the client has reported the RPE-AM, ask the client to also assess the RPE-O.
- Remind the client during the repetition *preceding* each RPE measurement, "Think about your feelings of exertion."
- Ensure that the RPE scale is in clear view of the client at all times.

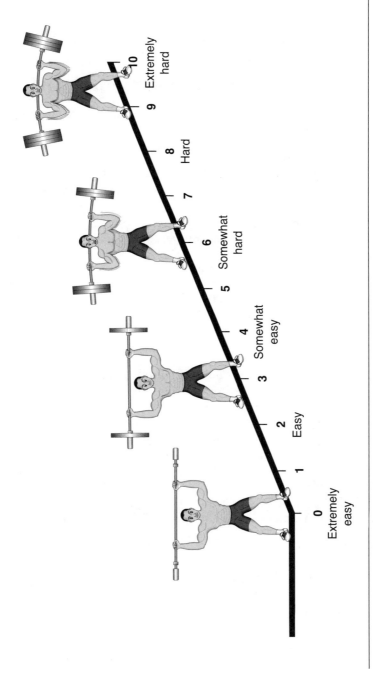

Figure 4.7 OMNI Resistance Exercise Scale for perceived exertion.

OMNI scale RPE responses during resistance exercise can be determined before and immediately after the client has participated in a weight training program. If the RPE for a given exercise weight is lower in the post-training test, the client's muscular strength has improved.

SPECIAL TESTING CONSIDERATIONS

The following is a list of helpful tips related to measuring RPE during exercise testing (Noble and Robertson 1996).

Perceptual Preference
- Cycle ergometer pedal rates between 50 and 80 rpm are metabolically efficient and perceptually preferable, meaning that they have the lowest RPE.
- Clients can preferentially change the speed increment from one treadmill stage to the next without adversely affecting measures of $\dot{V}O_2$max and HR.

Medications
- RPE generally is not affected by drugs such as propranolol and atropine. However, these drugs do, respectively, decrease and increase HR.

Muscle Mass
- When a client performs at a given submaximal $\dot{V}O_2$ or PO, RPE is often higher for those test modes that require the use of a smaller muscle mass (such as the arms) than a larger muscle mass (such as the legs).
- However, RPE usually is similar in small muscle mass and larger muscle mass exercise when compared at a specified % $\dot{V}O_2$max.

Test Termination
- The RPE should be at or very near the highest number on the rating scale when the client stops a maximal GXT due to fatigue.

Effect of Age
- RPE does not differ substantially between younger and older adults if factors such as aerobic fitness, muscular strength, and daily physical activity are the same.
- In general, the RPE at a constant aerobic or resistance exercise intensity increases as a client's level of aerobic fitness and muscular strength decline with age.

Sleep Loss

- In general, if the client has lost substantial sleep over a 48 to 72 h period preceding a GXT, the RPE responses at submaximal stages or intensities are elevated.

Psychological Factors

- Clients who are clinically depressed or anxious may demonstrate greater day-to-day fluctuations in their RPEs.

Effect of Gender

- The menstrual cycle does not appear to affect RPE responses during a GXT.
- The possible effect of gender on RPE is described in chapter 7.

Ambient Setting

- Factors such as background music and room lighting have not been shown to consistently influence RPE.

SUMMARY

In both health-fitness and clinical settings, a simple-to-use, subjective guide for determining physiologically appropriate and safe aerobic and resistance exercise intensities is needed. RPE-based exercise tests meet this need. RPE-based tests provide practitioners with information about the accuracy of the client's exercise prescription and whether adjustment of the training dosage is required. In addition, these assessments provide feedback to the client about training progress and fitness gains. The results of these tests can be used to estimate fitness, prescribe training programs, and track conditioning progress. Using RPE as a criterion test measure has several advantages: the assessments employ a submaximal end point, are comparatively quick to administer, and require minimal equipment and instrumentation.

C·H·A·P·T·E·R

5

Exercise Programs Using a Target Rating of Perceived Exertion

CASE STUDY

Client Characteristics

The client is a 55-year-old male who has a sedentary occupation and has not regularly exercised in approximately 20 years. During a recent medical examination, the following cardiovascular risk factors were identified: excess body fat, elevated cholesterol, a low high-density lipoprotein level, and a poor aerobic fitness level.

Exercise Need

The client's physician recommended structured and supervised aerobic exercise as part of a comprehensive lifestyle modification program to lower the client's cardiovascular disease risk and improve his overall quality of life. Consultation with an exercise practitioner was recommended as the first step in developing an individually prescribed aerobic training program.

Action Plan

The practitioner elected to develop an aerobic exercise prescription using a combined RPE estimation and production procedure based on the client's cardiovascular risk profile. First, a target RPE zone of 12 to 14 (Borg scale) was determined from perceptual estimates made during the preparticipation GXT. The client was then instructed to adjust his training intensity during each exercise session to produce the previously estimated target RPE. Training at a target RPE derived from GXT responses provides an overload training stimulus that is both physiologically effective and medically safe.

T he objective of an aerobic exercise prescription is to identify the training intensity that requires a **threshold** level of oxygen consumption (Robertson 2001b). When the threshold is reached, the training intensity produces an overload stimulus that improves aerobic fitness. Exercise performed at a **target RPE**—the level of exertion to be achieved—leads the client to achieve and maintain this physiological overload during an aerobic training session. When developing an exercise prescription, the practitioner determines the target RPE during a pretraining GXT. The target RPE is then used to guide training intensity according to the client's level of aerobic fitness and activity preference. When the client achieves the target RPE during a given training session, the HR, $\dot{V}O_2$, rate-pressure product, and electrocardiographic responses will be the same as those observed at the same RPE during the GXT (Robertson and Noble 1997).

The target RPE for aerobic exercise training is determined by using either a combined estimation and production procedure or a production-only procedure (Noble and Robertson 1996), depending on the aerobic fitness level or medical status of the client. If the client is sedentary, overweight, or has medical limitations such as cardiopulmonary or orthopedic problems, then a combined RPE estimation and production procedure should be used. In contrast, if the client regularly participates in aerobic exercise and does not require close supervision, an RPE production-only prescription procedure is both effective and safe. This chapter focuses on the use of a target RPE derived and applied using an **estimation–production prescription procedure.** Chapter 6 describes exercise prescription using primarily an RPE **production-only procedure.**

ESTIMATION–PRODUCTION PRESCRIPTION PROCEDURE

Developing a prescription for an aerobic exercise program using a perceived-exertion estimation–production procedure involves two steps. First, a target RPE is determined using RPEs that were *estimated* by the client during a GXT. Second, the client then *produces* this target RPE by continuously adjusting the exercise intensity during each training session. When exercise training is performed at an intensity that produces the target RPE, the client has achieved an overload stimulus that will improve aerobic fitness.

RPE Estimation Procedure

1. Measure RPE and $\dot{V}O_2$ during each stage of the preparticipation GXT.

2. Plot the RPE for each exercise stage against its corresponding $\dot{V}O_2$.

3. Draw a reference line representing the best fit through the data points.

4. Determine the client's actual or estimated $\dot{V}O_2$max according to the measurements made during the last stage of the GXT. *(Note: $\dot{V}O_2$max is a general term denoting the maximal amount of oxygen that can be consumed during treadmill exercise while breathing ambient air at sea level. $\dot{V}O_2$peak describes this value obtained during cycle and arm-crank ergometry. Because both $\dot{V}O_2$max and $\dot{V}O_2$peak can be influenced by clinical responses during exercise testing, the term symptom-limited $\dot{V}O_2$max/peak can also be used.)*

5. Calculate the overload training zone—normally, 70 to 85% of the $\dot{V}O_2$max. Alternatively, it can be calculated as 50 to 85% of the $\dot{V}O_2$ reserve, that is the percent range between resting and maximal $\dot{V}O_2$ (ACSM 2000).

6. Draw vertical lines from the $\dot{V}O_2$ values on the x-axis corresponding to 70 and 85% of the training zone to the reference line. See figure 5.1 for an example.

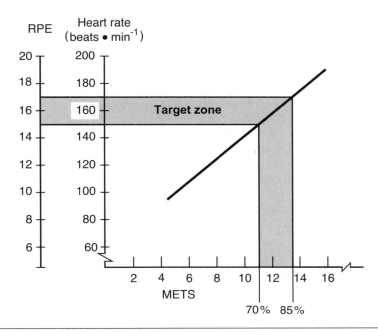

Figure 5.1   Prescription of target RPE training zones using an estimation–production procedure.

7. Draw horizontal lines from the points of intersection on the reference line to the y-axis.

8. The intersection points on the y-axis mark the low and high target RPEs to be used during training. These target RPEs demarcate the overload training zone that will improve aerobic fitness.

Producing the Target RPE

Once the client's target RPE zone has been identified with the GXT responses, it can be used to guide the intensity of the aerobic training program. That is, the client is directed to produce during each exercise session the target RPE (or target RPE range) that was estimated during the GXT to provide the optimal overload training stimulus to improve aerobic fitness. The client adjusts the exercise intensity (that is, the speed of walking, running, stepping, cycling, swimming, or arm cranking) so that her level of perceived exertion is within the target RPE zone. Periodically remind the client that the speed of the exercise can be increased or decreased at any time in order to keep the level of exertion within the target RPE training zone.

Of practical importance is that the client will automatically adjust the training speed needed to achieve the target RPE as the level of aerobic fitness changes. When his aerobic fitness level increases over the course of the training program, the client will subconsciously increase the intensity of training to keep the target RPE within the prescribed zone. The physiological overload that occurs with the new and higher exercise intensity is correspondingly greater, and so the optimal training stimulus to improve aerobic fitness is continually produced.

The procedure for guiding the training intensity using a target RPE is similar to commonly used prescription procedures that employ a target HR. However, the target RPE method has a number of practical and time-saving advantages over a target HR. First, monitoring the HR by palpation can give inaccurate results and usually necessitates stopping exercise to count. Electronic HR monitoring with wristwatchlike devices can also disrupt the prescribed flow of the exercise program. Whatever the monitoring method, HR is influenced by air temperature, psychological stress, caffeine, and medications. Unfortunately, some clients become "pulse counters" and get less pleasure from the exercise program. In contrast, the subconscious process of adjusting the exercise intensity to reach a target RPE training zone has none of these functional problems.

Table 5.1 shows a sample aerobic exercise training session in which the intensity is guided by target RPEs. The client is a 45-year-old woman who has normal body weight and has used the cardiovascular workout equipment at her health club regularly for the past 2 years.

Table 5.1   Sample Aerobic Exercise Training Session

	Workout component		
	Warm-up*	**Stimulus****	**Cool-down***
Duration (min)	3–5	20–30	5

Target RPE
  Borg 6–20
  OMNI
	
	7–10
	1–3
		
		12–15
		5–7
			
			10–7
			3–1

* Includes stretching or flexibility movements and aerobic exercise; ** Aerobic exercise.

Note: Adjust the target RPE for the stimulus component downward by 2 Borg categories or 1 OMNI category for new exercisers (Borg 1998).

TARGET RATING OF PERCEIVED EXERTION ZONES FOR CROSS-TRAINING

Exercise conditioning programs in health-fitness and clinical settings often combine several different aerobic training modes in the same exercise session. Called *cross-training,* these workouts are for total body conditioning and add variety to the exercise program, which prevents boredom and promotes program adherence. In cross-training, a number of different aerobic activities are performed separately in a particular sequence. Normally, cross-training intensity is set by target HRs that were determined from separate GXTs for each of the training modes. However, multimodal exercise testing is often impractical and expensive, especially when a comparatively large number of different exercises are involved. An alternative approach is to use a single target RPE to guide the intensity of each of the individual exercises as the client moves from one training mode to the next. This is called a **cross-modal target RPE.** Robertson and colleagues (1990a) validated its use in regulating aerobic training intensity by demonstrating that a cross-modal target RPE derived from a treadmill test could be used to guide the intensity of stationary cycling and bench stepping with light hand weights. The latter activity combines arm and leg movements in a unique form of low-impact, total-body exercise (Auble, Schwartz, and Robertson 1987). When they were compared at an intensity equivalent to 70% of $\dot{V}O_2$max, it was found that the target RPE-O was the same for the treadmill GXT (the estimation test) and the cycling and stepping exercises (the production sessions). These findings indicate that a target RPE is physiologically valid for use during a cross-training program. Of note is that the target RPE derived from a treadmill GXT was equally effective in guiding the training intensity of the leg-only cycle exercise and the combined arm and leg hand-weighted bench stepping. Therefore, a single cross-modal target RPE can be used for both weight-bearing and non-weight-bearing aerobic exercises involving two or four limbs and performed in any order.

WALKING AND RUNNING EXCHANGES

An exercise prescription that alternates walking and running modes in the same training session adds flexibility to the workout. The same target RPE is used for both modes and should be identified using the procedures described earlier in this chapter. To determine walking and running exchanges, start by identifying the RPE intersection point, or RPEpt (Noble and Robertson 1996). Defined as the speed at which the perceived exertion is the same for both walking and running, the RPEpt is approximately 4.0 mph for healthy clients. At speeds below the RPEpt, walking requires less effort than running. At speeds above the RPEpt, walking requires more effort than running. Therefore, the training speed required to produce a target RPE will be slower for running than walking if the prescribed intensity is below the RPEpt of 4.0 mph. When the prescribed intensity is above the RPEpt, the training speed will be faster for running than walking. Figure 5.2 presents a typical exercise prescription that employs walking and running exchanges. A horizontal line is drawn from the target RPE on the y-axis so that it intersects both the walking and running curves. Vertical lines are then drawn to the x-axis from the points of intersection on the walking and running curves. The client produces the target RPE by selecting either a comparatively faster walking speed or a slower running speed.

Walking and running exchanges that are guided by a target RPE add flexibility to daily training sessions. The client can choose the mode for a particular session without compromising the physiological validity of the prescription. Accommodating the client's activity interests helps to promote long-term program adherence (Buckworth and Dishman 2002).

INTERMITTENT EXERCISE USING SLIDING TARGET RATINGS OF PERCEIVED EXERTION

When a training session includes intermittent exercise bouts, it is helpful to use *sliding target RPEs*. An intermittent format often allows the client to perform a greater amount of training in a single workout than is possible during continuous training, when accumulating fatigue may limit performance. Target RPEs at the lower and higher ends of the client's prescribed (% $\dot{V}O_2max$) training zone should be identified. For example,

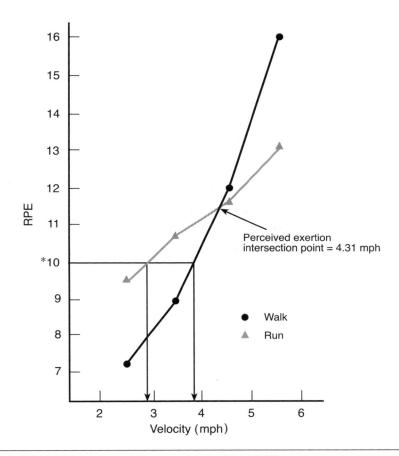

Figure 5.2   Walking and running exchanges at a target RPE.

Reprinted, by permission, from B.J. Noble et al., 1973, "Perceived exertion during walking and running-II," *Medicine and Science in Sports* 5: 116-120.

the client can exercise intermittently for 3 min at a target RPE of 13 on the Borg 6–20 Scale and then for 3 min at a target RPE of 17 (figure 5.3). Intermittent exercise at these sliding RPE intensities is then repeated for a prescribed number of times during the daily training session.

An intermittent training format that is guided by sliding target RPEs has important benefits for clients with low aerobic fitness levels and for patients with cardiopulmonary disease. Because certain physical activities require a higher energy level than can be tolerated for extended periods by clients who have low functional aerobic power, the client may be able to perform these activities only when an intermittent exercise format is used. In addition, because aerobic fitness can be limited by clinical

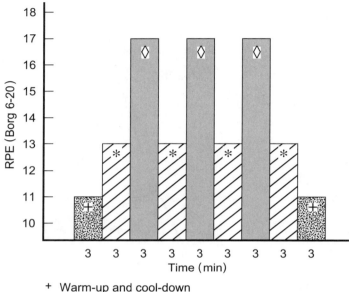

Figure 5.3   Sliding target RPE zones for aerobic exercise training.

symptoms, cardiac and pulmonary patients may not be able to sustain the prescribed exercise intensity long enough to produce a training stimulus. An intermittent format allows the client or patient to achieve the prescribed training stimulus during the higher intensity bouts, whereas the brief, intermittent lower intensity bouts help to minimize the clinical symptoms of exertional intolerance. The shortest possible low-intensity bout should be used and the period for the high-intensity bout should be shorter than that at which symptoms occur.

TARGET RATINGS OF PERCEIVED EXERTION: OVERALL VS. ACTIVE MUSCLE

For cross-training exercise prescriptions, a target RPE-O is more useful than a target RPE that is differentiated to the active limb muscles or to the chest. The intensity of the RPE for the legs and arms varies with the number of limbs that are involved. Limb involvement is determined by the exercise mode. As the exercise mode changes during the cross-training workout, so too will the number of limbs that are involved. Therefore, exclusive use of a target RPE for either the legs or the arms is not effective in guiding cross-training intensity. A good rule of thumb is that the target RPE-O is more useful than the target RPE-L or RPE-C for developing a total-body cross-training program.

A target RPE that focuses on the specific muscles that are activated during an exercise (RPE-AM) is particularly useful for guiding intensity in a resistance exercise training program. The muscle mass that is activated depends on the type of resistance exercise and, in some cases, whether the exercise is performed using free or machine weights. Using a target RPE for the feelings of exertion that arise from the active muscle groups is the most accurate method for guiding resistance training intensity, so the RPE-AM should be used to guide the target training zone.

A target RPE that is specific to exertional feelings arising in the chest is most helpful in clients with breathing difficulties such as emphysema and bronchitis. Linking the target RPE-C with feelings of labored breathing (dyspnea) helps to ensure that the intensity of the exercise program stays within medically safe limits, regardless of whether the arms only, legs only, or a combination of limbs are used.

TARGET HEART RATE HELPER

For the inexperienced exerciser, training intensity initially can be guided by both a target RPE and a target HR (Noble and Robertson 1996). The HR serves as a "helper" to the target RPE in guiding the intensity of the client's exertion. Determine the target HR when you calculate the target RPE zone as described earlier in this chapter.

When using this technique, instruct the client to adjust the training pace so that both the RPE and HR fall within their respective target zones. As the training program progresses, the need for HR monitoring lessens and eventually it can be eliminated. The training intensity can then be guided by the target RPE zone only.

PROBLEMS AND SOLUTIONS

The length of time that elapses between the identification of a target RPE during the preparticipation GXT and its use during training should be as short as possible (Van Den Burg and Ceci 1986). When there is a substantial delay (30 days or more) between initial testing and the first training session, the energy demand at the target training RPE may be greater than expected if the client's memory of how the target RPE feels has faded, resulting in too-intensive exercise and thus greater $\dot{V}O_2$ and HR response. The solution to this problem is to repeat the GXT to reestablish the target RPE.

A similar situation occurs when the PO or speed that is associated with a target RPE is greater during a training session than it had been during the preparticipation GXT. This results from the greater length of time typically required for an individual training session ($\geq$30 min) compared to the 12 min needed to administer a GXT (Noble and Robertson 1996). However, this is not really a problem. The purpose of the exercise prescription is to use the target RPE to achieve a given $\dot{V}O_2$ level. The prescription does not ask the client to achieve a preestablished PO or exercise speed. Rather, the client is encouraged to *change* the PO or speed to achieve the target RPE. Therefore, it is expected that clients will have a PO or exercise speed that differs between testing and training for a particular target RPE.

Occasionally, the HR response at a given target RPE during a training session is not the same as that noted during the GXT (Davies and Sargeant 1979). Such inconsistencies in HR responses may be attributable to multistage testing protocols in which the exercise intensity is increased with each stage. The cumulative physiological stress of the preceding stages raises the HR during exercise testing above the level reached during training. Prescription errors of this type can be minimized by closely monitoring the client's HR during the first several training days. Determine the amount by which the training HR is above the expected value based on the GXT responses. Return to the original prescription data and replot the target RPE downward until it matches the actual HR that the client has consistently achieved during training. This procedure, known as *client calibration*, can be used on those comparatively rare occasions when the HR differs between testing and training.

SUMMARY

This chapter explains the use of target RPE zones to guide aerobic training intensity. If the client is sedentary, overweight, or has medical limitations, then a combined RPE estimation and production procedure should be used to prescribe and regulate training intensity. In contrast, if the client regularly and independently performs aerobic exercise, an RPE production-only prescription procedure is appropriate. Examples of how target RPE zones are used to guide constant, intermittent, and cross-training programs are presented. It is recommended that the RPE-O be used as the target training zone for most types of aerobic exercise programs. Walking and running exchanges that use the same target RPE and provide program flexibility are discussed.

6

OMNI Rating of Perceived Exertion Zones for Health Fitness and Weight Loss

CASE STUDY

Client Characteristics

The client is a 60-year-old female who has normal body weight and no cardiovascular or orthopedic exercise limitations. She is an experienced exerciser who runs 4 miles on 3 to 5 days of each week and performs resistance exercise for 60 min on 2 days of each week. She has very high levels of both aerobic fitness and muscular strength and endurance.

Exercise Need

Because it is difficult to exercise outdoors in inclement whether, the client has requested that the practitioner design a conditioning program that (a) can be performed indoors, (b) provides a balanced aerobic and resistance exercise regime, (c) has variety that "breaks the normal routine," and (d) provides a sufficient overload stimulus to simultaneously improve aerobic and strength fitness.

Action Plan

The practitioner recommends that the client use a sliding RPE zone system to self-regulate the intensity of a combination training program that includes a mixture of aerobic and resistance exercise sets. The "combo" exercise sets are performed sequentially, with the aerobic and resistance exercises undertaken at alternating stations. The intensity of each aerobic set is adjusted to slide between an RPE of 5 and 7 (OMNI scale) so that it spans the anaerobic threshold. The resistance exercise sets are designed to increase local muscular endurance by starting at an RPE zone of 3 (OMNI scale) and progressing to a final RPE of 10. Both warm-up and cool-down sets are included, as is a stretch-band breakout station.

This chapter introduces the RPE zone system for training and weight loss, which has applications for health-fitness clients as well as selected patients who require exercise rehabilitation.

HEALTH-FITNESS GOALS IN THE CLUB AND CLINIC

For a health-fitness program to be effective, the client must achieve certain goals and the practitioner must meet certain professional obligations. First, it should be determined what the client expects from the training program. In today's fast-paced society, many clients lead busy lives, and family, work, and social obligations often compete with the limited time that is available for fitness activities. Fitting an exercise program into such a busy lifestyle can be challenging for both client and practitioner. Yet, the sedentary nature of our modern, highly automated society demands participation in health-fitness activities on a regular basis to promote health and general well-being. Therefore, the health-fitness program must deliver the maximum benefits in the shortest training time. In other words, it should efficiently accomplish the following health-fitness goals:

- Increase cardiovascular fitness, muscular strength and endurance, and flexibility
- Provide help with reaching and maintaining ideal body weight
- Be pleasurable by including preferred activities that require a reasonable but not excessive time commitment

Health-fitness practitioners are obligated to help their clients meet their training goals. They must provide their clients with safe and effective exercise programs while simultaneously meeting the financial requirements of the club or exercise facility. The goal is to efficiently employ the facility's dual resources of time and space. Normally, the time allotted for the use of a given exercise space is scheduled in advance, and it is expected that the space will be fully used throughout the facility's operating hours. Therefore, it is essential for the client and practitioner to maximize their use of the time allotted for a given training space. Daily orientation, warm-up, stimulus exercising, and cool-down should be undertaken smoothly and in a timely manner, but without sacrificing training outcomes. The exercise environment should promote adherence to the training program by

- encouraging the client to exercise at the same time each day and on the same days each week;

- including the client's friends and acquaintances as part of the exercise group;
- providing computer-using clients with regular progress reports via e-mail or Web page; and
- including in the exercise program new and technologically innovative exercise machines, preferably those that have interactive computer controls.

This chapter describes how RPE training zones can be used to meet the goals of health-fitness clients and the professional obligations of exercise practitioners.

WHAT IS A RATING OF PERCEIVED EXERTION TRAINING ZONE?

A target RPE training zone employs a target RPE range that (a) identifies the exercise intensities that will improve cardiovascular fitness, reduce excess body weight, and increase muscle strength and (b) is appropriate for clients who have certain characteristics in common, such as age, fitness level, gender, and training experience. When clients exercise within a prescribed RPE zone, they work at the appropriate intensity to achieve their personal training goals. A unique aspect of the RPE zone system is that it does not require extensive preparticipation testing to calculate the client's target training intensity. Instead, the training intensity is dictated by the RPE zone that is appropriate for the characteristics of the client's group. It is important to note that because the RPE zone system does not require preparticipation assessment, it is primarily recommended for use by clinically normal clients in whom medical risk is not an overriding consideration in determining the appropriate exercise training intensity (Robertson 2001b).

RPE Conversion

The RPE training zones described in this chapter can be used with the OMNI scale format or the two most common Borg scales, the 6–20 scale and the CR-10. To help the practitioner translate the zones from one scale to another, an RPE zone conversion chart is presented in figure 6.1. This allows the practitioner to prescribe a target training zone using one type of RPE scale and convert it to one of the other two scales without sacrificing training accuracy. However, once a particular scale format has been selected, the client should rely primarily on that scale to guide the exercise training zones.

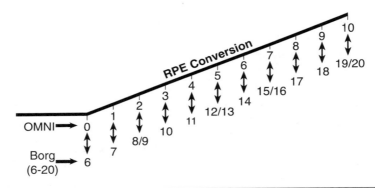

Figure 6.1   RPE conversions between the OMNI, Borg 6–20, and Borg CR-10 scales. The OMNI and Borg CR-10 scale units are equivalent.

Anaerobic Threshold Zone

One of the most effective RPE zones for an aerobic training program has an intensity equivalent to the client's anaerobic threshold (Robertson et al. 2001). The **anaerobic threshold** is the exercise intensity at which the blood's lactic acid concentration begins to increase above resting levels. For most children and adults, the exercise intensity that equals the anaerobic threshold is approximately 50 to 80% of their $\dot{V}O_2$max. When clients exercise in the RPE zone equivalent to their anaerobic threshold, they have an appropriate overload stimulus to improve their aerobic fitness and reduce excess body weight. For most adults, the RPE zone that spans the anaerobic threshold is 5 to 7 on the OMNI scale and 12 to 14 on the Borg 6–20 Scale. The client is instructed to alter the exercise intensity (pace) until his RPE is within the anaerobic threshold zone. Once the client has reached this RPE zone, he simply alters the intensity as needed to remain in the training target zone. This comparatively simple procedure is used throughout the conditioning session. To help clients identify and stay within their training RPE, the target zone on the OMNI scale can be shaded, as shown in figure 6.2.

Sliding RPE Zones

In a sliding zone exercise program, two or more target RPE zones are used in the same workout (figure 6.3). The client begins in the *warm-up zone* and then increases the exercise intensity until her RPE reaches the *first stimulus zone.* Exercise is performed in this zone for 2 to 3 min. Next, the client increases, or "slides up," the exercise intensity to the *second stimulus zone* and maintains it at that level for 2 to 3 min. At the end of the second-zone exercise period, the client decreases, or "slides down," the intensity

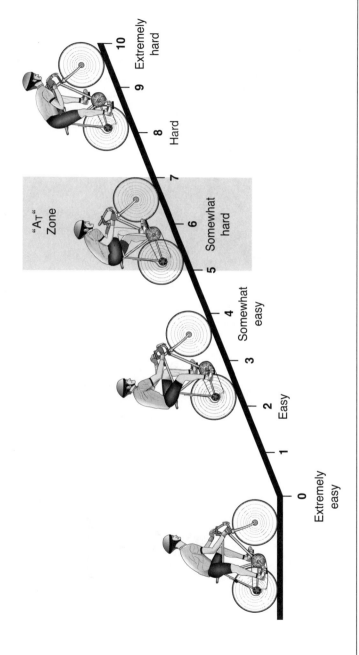

Figure 6.2 The RPE training zone equivalent to the anaerobic threshold using the OMNI cycle scale.

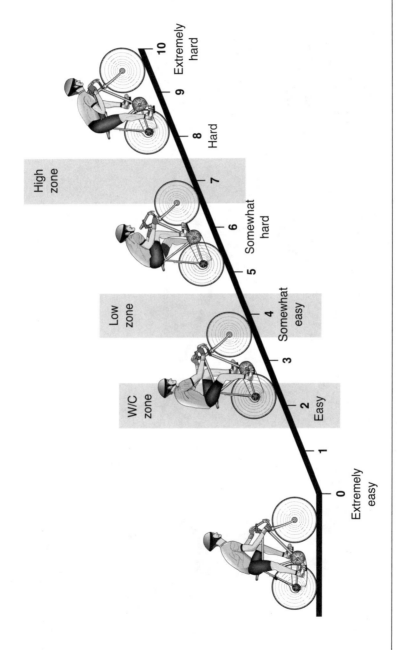

Figure 6.3 Sliding RPE training zones using the OMNI cycle scale. W/C = warm-up and cool-down.

so that the RPE is again in the first stimulus zone. Sliding can be done as many times as desired. When the stimulus portion of the exercise session has been completed, the client slides down the intensity to the *cool-down zone,* at which the exercise is performed for 1 to 3 min. The sliding zone system varies the training routine, and clients often report that it is fun to slide the intensity up and down. When clients view exercise as fun and pleasurable, they are more likely to adhere to a regular training schedule. Several of our clients at the University of Pittsburgh's CEHFR refer to sliding between zones according to the prescribed RPE as *R slides.*

RPE Climbs to Determine Exercise Duration

The duration of intermittent training bouts can be guided by *RPE climbs.* This system takes into account that at a set exercise intensity, RPE increases as the intermittent training bout gets longer. The RPE climbs technique follows this procedure:

1. Identify the starting intensity (speed or pace) needed to produce the target RPE on the OMNI scale.
2. Have the client hold the intensity constant during the training bout.
3. Ask the client to estimate the OMNI RPE every 3 to 5 min.
4. When the RPE climbs 2 units above the target, stop the exercise bout and let the client rest.
5. Have the client begin the next exercise bout at the original speed and target RPE and repeat steps 2 through 4.
6. Direct the client to continue using the intermittent RPE climbs technique until the stimulus portion of the workout session is over.
7. As fitness improves, it will take the client longer to exceed the target RPE by 2 units.

USING RATING OF PERCEIVED EXERTION ZONES FOR FITNESS AND WEIGHT LOSS

The following sections demonstrate how clients can use RPE zones in various exercise situations. Information pertinent to clients of all ages, abilities, and fitness levels is included for health-fitness practitioners, at-home exercisers, and those who prefer aquatic exercise.

RPE Zone System for Cross-Training

As stated previously, the RPE zone equivalent to the anaerobic threshold is the same for a wide range of aerobic exercises, including walking, running, cycling, stepping, arm cranking, sliding, and rowing. Therefore, clients can use the anaerobic threshold zone to guide the intensity of a cross-training program in which a number of different types of aerobic exercises are performed in a single workout session. The client simply proceeds through the cross-training sequence, adjusting the intensity of each exercise by producing the target RPE zone equivalent to the anaerobic threshold. Conveniently, the same RPE zone is used for leg exercise on a treadmill, arm exercise using an arm crank, and combined arm and leg exercise performed when rowing or climbing. Therefore, different types of exercise that involve both the upper and lower body can be undertaken in the same workout. This multimodal application of the RPE zone system is useful in children as well as adults (Duncan et al. 1996). As the client's aerobic fitness improves, she automatically adjusts the intensity required to produce the target RPE zone (the anaerobic threshold) for each exercise in the cross-training program.

RPE Training Zones for Children and Youth

The RPE zone equivalent to the anaerobic threshold can also be used to guide training programs for children. For girls and boys 8 years of age and older, the RPE zone that is equivalent to the anaerobic threshold ranges from 5 to 7 on the children's OMNI scale (Robertson et al. 2001) and 12 to 14 on the Borg 6–20 Scale (Mahon, Gay, and Stolen 1998). Children often prefer intermittent free-form exercise with alternating intensity levels. The sliding RPE zone system works very well in helping young children to achieve an exercise intensity that makes the activity fun while at the same time promoting optimal levels of aerobic fitness and body weight. Research at the University of Pittsburgh's CEHFR has shown that children are quite adept at sliding their exercise intensity from a low to a moderate and back to a low target RPE zone (Robertson et al. 2002). The OMNI picture cues that depict various types of play help children perform within the RPE training zone that is equivalent to their anaerobic threshold.

Training Improvements and Daily Adjustments

One advantage of using the RPE zone system is that clients automatically adjust it to suit changes in their aerobic fitness. That is, as their level of aerobic fitness increases during training, clients automatically increase the exercise intensity to reach their target RPE zone. Similarly, if clients have an "off" day because of a lingering head cold or modest fatigue, they automatically adjust the intensity downward, although they still reach the target RPE zone for that training day.

Sliding Zone System for Resistance Exercise Training

Resistance exercise training for health-fitness and competitive athletics typically employs either an **intensity-loading protocol** or a **volume-loading protocol.** When intensity loading is used, the % 1RM is progressively increased from one exercise set to the next. Remember that the 1RM is the maximum weight that can be lifted one time and that it is different for each resistance exercise. In general, the higher the % 1RM, the fewer the number of repetitions for each set. When a volume-loading protocol is used, the number of repetitions increases progressively from one exercise set to the next while the intensity (the % 1RM) does not change (Robertson 2003). A key word used in describing both of these protocols is **progressive.** During resistance exercise, the training stimulus progresses until a predetermined end point is reached for each set. For most clients and athletes, this end point is at the moment when they are unable to perform one additional repetition with a given weight, termed *muscular failure.* However, some clients, particularly those who are sedentary and overweight, may need an end point at less than muscular failure. This lower end point is especially helpful during the early phase of the strength training program. The resistance exercise stimulus for each set should always progress systematically until the designated end point is achieved.

The goal of resistance exercise training is to improve muscular strength and local muscular endurance and to stimulate muscular **hypertrophy** (an increase in size). All three of these training outcomes require a progressive resistance stimulus that can be achieved with intensity or volume loading. A muscular endurance program employs comparatively more repetitions and lighter weights (the % 1RM) for each exercise set. In contrast, a muscular strength program employs fewer repetitions and heavier weights for each set. When an increase in muscular size is the training goal, the number of repetitions and the weight that is lifted roughly fall between what is required for developing muscular endurance and what is required for muscular strength.

The sliding RPE zone system is ideally suited to providing a progressive training stimulus that promotes muscular strength, endurance, and hypertrophy. The OMNI-RES sliding zone system is shown in figure 6.4. The examples of this system that are discussed below may help practitioners understand how to develop an individualized resistance training program for their clients.

In the first example, the client's training goal is to increase her muscular endurance.

- First, select the exercise to be used, such as a biceps curl or knee extension.

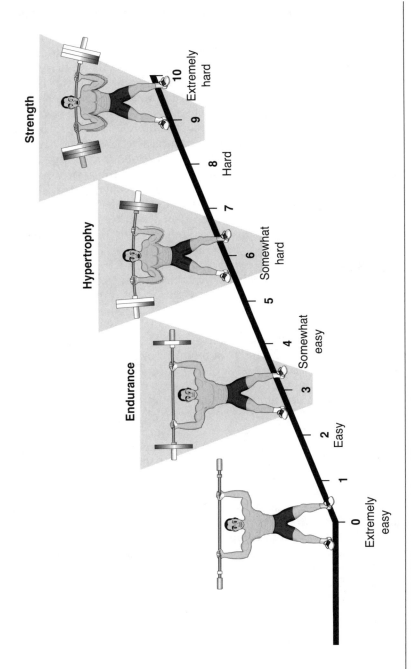

Figure 6.4 OMNI RPE training zones for resistance exercise.

- Have the client use trial and error to identify what weight feels like an RPE of 3 on the OMNI-RES. The client will likely need to lift several different weights until she identifies the one that produces exertion equal to an RPE of 3. Explain this trial and error procedure to the client during the pretraining orientation period. Over time, the client will become almost automatically able to select the appropriate starting resistance for any given exercise.

- Next, have the client begin the exercise set by lifting and lowering the RPE 3 weight. The lifting (concentric) and lowering (eccentric) phases of the exercise should each be completed in 2 s. Each complete lifting and lowering cycle is a single repetition.

- The level of exertion increases as the number of completed repetitions increases.

- When the RPE has reached 10, no further repetitions can be made and the set is finished.

- As needed, the client can perform additional sets of the same exercise with the RPE again progressing from 3 to 10 on the OMNI-RES. Allow an appropriate rest period between sets.

- Be sure that the OMNI-RES showing the sliding zones is in full view of the client at all times.

In the next example, the sliding RPE zone system is used to improve muscular strength.

- Have the client select a training weight that produces an RPE of 9 on the OMNI-RES for the resistance exercise that is to be performed (see figure 6.4).

- Have the client perform the exercise repetitions until the RPE has reached a 10, at which point no further repetitions are possible and the set is complete.

- Additional sets can be performed using the same RPE sliding zone progression from 9 to 10.

A sliding zone program to promote muscular hypertrophy uses a training weight that the client selected to produce an RPE of 6 on the OMNI-RES. The RPE zone progresses upward in the manner described for the strength and endurance programs until the set is complete.

Note that the sliding zone system does not use a predetermined number of repetitions for each set. The client automatically determines the number of repetitions required for her RPE to progress from 3 to 10. In this way, the sliding zone accommodates the client's level of strength and fatigue by letting the client control both the amount of weight being lifted and the number of times it is lifted. In effect, the client performs a training program of her own design, which ultimately increases training compliance.

One important advantage of the sliding zone system is that improvements made by the client during resistance exercise training are automatically adjusted for. As the level of muscular strength (or endurance) increases, so too does the amount of weight that the client needs to lift to reach the target RPE zone. In this way, zone progression always provides the optimal training stimulus for improvement.

Table 6.1 presents general guidelines for resistance exercise training using OMNI RPE zones for healthy adults and cardiac patients.

Table 6.1 Protocols and RPE Zones for Resistance Exercise Training

Client	OMNI RPE zone	Number of sets	Maximum number of repetitions	Number of exercises	Frequency (days/week)
Healthy sedentary adult	3	1	8–12	8–10	2–3
Healthy sedentary elderly adult	2–3	1	10–15	8–10	2–3
Cardiac patient (low risk)	2–3	1	10–15	8–10	2–3

Adapted from M. Feigenbaum and M. Pollock, 1999, "Prescription of resistance training for health and disease," *Medicine and Science in Sports and Exercise* 31(1): 38-45.

Resistance Exercise for the Wheelchair Client

Resistance exercise programs that employ target RPE training zones can be very effective for clients who use a wheelchair. These resistance programs use weights or stretch bands to strengthen the back, chest, shoulders, biceps, and triceps muscles. If wheelchair clients can move their legs, they can also perform seated leg extensions and isometric leg strength exercises. The resistance exercise intensity for wheelchair clients can be guided by RPE zones using the OMNI-RES. Wheelchair clients generally benefit most from using RPE zones that improve both muscular endurance and strength. Table 6.2 shows a sample resistance program that uses weights or stretch bands.

Table 6.2 Resistance Exercise for the Wheelchair Client

	Endurance RPE	Strength RPE	Number of sets
Weights			
Military press	3	9	2–3
Biceps curl	3	9	2–3
Diagonal or front raise	3	9	2–3
Triceps kickback	3	9	2–3
Stretch band			
Latissimus pull-down	3	9	2–3
Chest fly	3	9	2–3
Triceps extension	3	9	2–3
Shoulder Internal rotation External rotation	 3 3	 9 9	 2–3 2–3

From S. McKechnic, 2002, "Fitness meets special needs," IDEA Personal Trainer. Reproduced with permission of IDEA Health & Fitness Association, (800) 999-IDEA, www.IDEAfit.com.

RPE training zones for wheelchair resistance exercise programs are employed as follows:

- Identify the muscle groups that are to be trained.
- Determine the OMNI RPE zone for that workout session: zone 3 for endurance, zone 9 for strength.
- Have the client select, by trial and error, the weight or stretch band that produces the target RPE zone.
- Direct the client to begin exercising.
- Exercise repetition should continue until the client reaches an OMNI RPE of 10, when the set is complete.
- Move to the next resistance exercise.
- Remember that one exercise repetition is a complete contraction and relaxation of the target muscle group against the resistance. The OMNI scale should be in full view of the client at all times.

RPE Zone System for Combination Training

The sliding zone system is ideally suited for combination training that involves a mixture of aerobic and resistance exercise sets. We have seen how the target RPE zone can be used to guide aerobic exercise intensity so that it is equivalent to the anaerobic threshold and to regulate repetitions during resistance exercise training. These two types of exercise can also be included in a combination training program that uses the RPE zone system to guide conditioning intensity.

- As shown in the program plan in figure 6.5, the client begins by performing a cycle exercise set at an intensity that produces an RPE within the warm-up zone on the OMNI cycle scale.
- The client then moves to a resistance exercise set, during which he produces an RPE equivalent to the warm-up zone on the OMNI-RES.

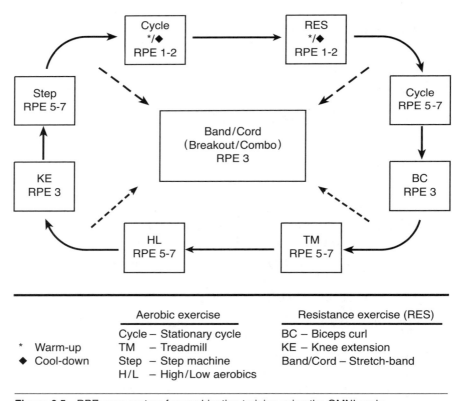

	Aerobic exercise	Resistance exercise (RES)
	Cycle – Stationary cycle	BC – Biceps curl
* Warm-up	TM – Treadmill	KE – Knee extension
◆ Cool-down	Step – Step machine	Band/Cord – Stretch-band
	H/L – High/Low aerobics	

Figure 6.5 RPE zone system for combination training using the OMNI scale.

- After completing the two warm-up sets, the client progresses sequentially through the remaining sets.
- The client adjusts the intensity of the aerobic sets to reach a target RPE zone of 5 to 7 on the OMNI scale, which is the anaerobic threshold.
- The purpose of the resistance sets is to increase local muscular endurance. Therefore, the client lifts the weight that produces an initial RPE of 3 and then repeats the exercise until he reaches an RPE of 10.
- The client can perform a "breakout" stretch-band exercise set for muscular endurance at an RPE of 3 at any time during the combo circuit.

Exercise continues until the client has completed the entire circuit. Remember that the final aerobic and resistance sets should be performed at RPEs that fall within their respective cool-down zones. An OMNI RPE scale with picture cues that match the exercise should be positioned in clear view of the client at each combo exercise station.

As a complement to the combination circuit in figure 6.5, table 6.3 presents a 10-week "cardioresistance" training format that allows clients to select the resistance and aerobic exercises that they prefer.

Table 6.3   Combined "Cardioresistance" Training Program Using OMNI RPE Zones

	Resistance exercise		Aerobic exercise	
Week	Number of sets	RPE zone	Time (min)*	RPE zone
1	1	3	12.5 (5 × 2.5)	5–6
2	2	3	20 (8 × 2.5)	5–6
3–4	3	6	37.5 (15 × 2.5)	5–6
5–8	3	6	37.5 (15 × 2.5)	6–7
9–10	3	6	37.5 (15 × 2.5)	7–8

* Numbers in parenthesis indicate the number of separate, 2.5 minute aerobic exercise bouts to be performed in a given training session. Resistance exercise sets should be interspersed between each aerobic exercise bout.

Adapted, by permission, from G. Sforzo et al., 1998, "Cardioresistance exercise: A new training technique," *ACSM's Health and Fitness Journal* 2(6): 11-17.

Indoor Group Cycling

Group cycle training—sometimes referred to as *spinning*—is performed in one of two structured and interchangeable formats, power spinning and speed spinning. Power spinning employs higher pedal resistance with slower pedal rates whereas speed spinning employs lower resistance and faster pedal rates. Group cycling sessions normally are performed under the direction of an exercise leader and often use music to set the pedaling cadence. Table 6.4 presents the format for a 50 min group cycling program that uses target RPE zones to guide spinning intensity. This program is appropriate for a group of clients who have mixed levels of fitness and cycling experience. The spinning program detailed in table 6.5 is for more advanced cycle trainers.

Table 6.4   Group Cycling Program Using RPE Zones

Time (min)	Description	% of maximum intensity	RPE zone	
			OMNI scale	Borg 6–20 Scale
0–5	Gradual, active warm-up	40–50%	4	11
6–10	Gradual hill	55–70%	6–7	14–16

11–19	Anaerobic interval training (6 sets):
	  Active interval: 30 s
	  Rest interval: 1 min
		≅85%
		60–70%
			8
			6–7
				17
				14–16

20–21	Active recovery (level grade)	50–60%	5–6	12–14

22–33	Aerobic interval training (2 sets):
	  Active interval: 3 min
	  Rest interval: 3 min
		75–85%
		50–60%
			7–8
			5–6
				15–17
				14–16

34–35	Active recovery	40–50%	4	11

36–38	Rolling hills (3 sets):
	  Active interval: 15 s
	  Rest interval: 45 s
		>95%
		50–60%
			9–10
			5–6
				18–20
				12–14

39–45	Level-grade endurance training	70–80%	7–8	15–17
46–50	Gradual, active cool-down	20–40%	1–2	7–9

From J. Smith, 2002, "Revisiting energy systems," IDEA Personal Trainer. Reproduced with permission of IDEA Health & Fitness Association, (800) 999-IDEA, www.IDEAfit.com.

Table 6.5   OMNI RPE Zones for a Studio Cycle Spinning Workout

Stage/min	OMNI RPE zone	Description	Feeling
1/1–4	2	Instruction and upper extremity warm-up	Easy
2/5–9	2	Lower extremity warm-up	Easy
3/10–13	4	Alternate sitting and standing in 45 s intervals with light resistance	Somewhat easy
4/14–16	8	Alternate seated sprints in 30 s intervals with 30 s slow cadence	Hard
5/17–20	10	Alternate standing sprints at a high resistance and high cadence in 20 s intervals with 20 s slow cadence	Extremely hard
6/21–25	7–8	Resistance intervals at constant intensity	Hard
7/26–28	8–9	Alternate seated, low-resistance, maximum-cadence 20 s intervals with 20 s at decreased cadence	Hard
8/29–32	10	Alternate sitting and standing on the third pedal revolution at maximum resistance	Extremely hard
9/33–36	8–9	Seated spinning with one final 20 s sprint	Hard
10/37–39	2	Cool-down, seated, at a decreasing cadence to a gradual stop	Easy

Adapted, by permission, from P. Fracis et al., "Physiological response to a typical studio cycling session," 1999, *ACSM's Health and Fitness Journal* 3(1): 30-36.

RPE Zones for Marathon Training

Tables 6.6, 6.7, and 6.8 present a 12-week training program for clients who are preparing to run a marathon. Separate programs are listed for novice and veteran runners. The RPE zones to guide training intensity during daily workouts are included.

Table 6.6   RPE Zones for a 12-Week Marathon Training Program

Experience level	Weeks	Workout distance (miles/day)	RPE zone OMNI	RPE zone Borg 6–20	Feeling
Novice	1–4	4–12	4–5	11–13	Somewhat easy to somewhat hard
	5–10	2–6	6–7	14–16	Somewhat hard to hard
		7–23	5–6	12–14	Somewhat easy to somewhat hard
	11–12	3–10	4–5	11–13	Somewhat easy to somewhat hard
Veteran	1–4	4–14	5	12–13	Somewhat easy to somewhat hard
	5–11	Tues (weeks 5, 6, 10), 4–10 repetitions, 440 yd	7–9	15–18	Hard
		Tues (weeks 7, 8, 9), 2–6 repetitions, 1,320 yd	6–7	14–16	Somewhat hard
		Thurs (weeks 3–6)	6	14	Somewhat hard
		Remaining days (any training distance)	5	12–13	Somewhat easy to somewhat hard
	12	Mon and Wed 3–6	6	14	Somewhat hard
		Tues and Thurs 8	5	12–13	Somewhat easy to somewhat hard

From IDEA Personal Trainer, Jan. 1999.

RPE Zones for Activities of Daily Living

It is estimated that only 22% of adult Americans regularly engage in activities that promote optimal health and fitness. This is the case despite growing scientific evidence that links daily participation in physical activities with improvements in both the quality and quantity of life (Goss et al. 2003). Part of the reluctance to become physically active may be related

Table 6.7　Twelve-Week Training Program for a Marathon: Novice

Week	Mon	Tues	Wed	Thurs	Fri	Sat	Sun	Total
				Miles				
1	0	6	0	4	6	0	4	20
2	0	6	0	4	6	0	6	22
3	0	7	0	4	6	0	9	26
4	0	8	0	4	6	0	12	30
5	4	0	8	6	0	15	0	33
6	5	0	8	6	0	17	0	36
7	6	3	5	6	0	16	0	36
8	3	4	8	6	0	19	0	40
9	2	6	5	4	0	23	0	40
10	4	0	8	6	0	15	0	33
11	6	0	10	3	0	8	0	27
12	6	8	3	8	0	0	Race day	

Modify the program according to the client's fitness level, running experience, goals, and response to the program. Approximately 10% of the week's total running time should include varied terrain, such as hills.

From S. Black, 1999, "Training clients to run a marathon," IDEA Personal Trainer. Reproduced with permission of IDEA Health & Fitness Association, (800) 999-IDEA, www.IDEAfit.com.

Table 6.8　Twelve-Week Training Program for a Marathon: Veteran

Week	Mon	Tues	Wed	Thurs	Fri	Sat	Sun	Total
				Miles				
1	0	6	8	4	6	4	12	40
2	0	7	9	4	6	4	12	42
3	0	7	10	4	6	4	13	44
4	0	8	10	4	7	5	14	48
5	9	6	12	6	5	15	0	53
6	8	6	14	6	5	17	0	56
7	8	8	11	6	5	19	0	57
8	8	8	12	6	5	21	0	60
9	8	8	10	6	4	23	0	59
10	5	6	10	6	3	18	0	48
11	6	0	10	3	0	8	0	27
12	6	8	3	8	0	0	Race day	

Modify the program according to the client's fitness level, running experience, goals, and response to the program. Approximately 10% of the week's total running time should include varied terrain, such as hills.

From S. Black, 1999, "Training clients to run a marathon," IDEA Personal Trainer. Reproduced with permission of IDEA Health & Fitness Association, (800) 999-IDEA, www.IDEAfit.com.

to a belief that relatively intense, prolonged exercise is necessary to realize health and fitness benefits. However, this is not the case; many lower intensity activities improve physical fitness and reduce excess body weight. These are called **activities of daily living,** or ADLs. ADLs typically are of a light to moderate intensity (such as brisk walking) and should total 30 min over the course of the day. This "exercise lite" approach to physical activity promotes important health benefits when pursued on a daily basis.

When clients begin an exercise program, many of them believe that if the training activities are to have health benefits, they must be undertaken at intensities that produce pain and discomfort or are perceived to be very hard. However, physical activities such as gardening, brisk walking, and ballroom dancing are of a light to moderate intensity, requiring only 40 to 60% of $\dot{V}O_2$max. The RPE zone that corresponds to these intensities ranges from 7 to 10 on the Borg 6–20 Scale and 1 to 4 on the OMNI scale. Called the *daily activity zone,* this range is appropriate for leg, arm, and combined arm and leg exercises (Goss et al. 2003). The energy expenditure that is linked to the daily activity RPE zone ranges from 4 to 8 kcal/min for men and 3 to 6 kcal/min for women. These energy requirements promote weight loss and improve fitness.

Stability Ball Exercises

Stability ball exercises are excellent for developing core strength in the erector spinae, multifidus, and quadratus lumborum muscles of the spine and the rectus abdominis, internal and external obliques, and transverses abdominis muscles of the abdomen (ACSM 2002). Because ball exercise programs have variety and flexibility, they are popular with clients of all ages and fitness needs. Although stability ball exercise also has dynamic properties, its basic component is holding a position with an isometric muscle contraction. The length of time for which each position is held can be regulated with R slides as follows:

- Select a stability ball exercise position.
- Once the client has assumed the position, ask her for her RPE on the OMNI scale.
- Instruct the client to estimate the RPE every 15 to 30 s while the ball position is held.
- When the RPE slides upward by 2 ratings, stop the exercise and move to the next stability ball position.
- The OMNI scale should be in full view of the client at all times.

Table 6.9 lists core and lower body stability ball exercises that can be guided by the RPE sliding system. These exercises are especially useful for older clients.

Table 6.9   RPE Slides for Core and Lower Body Stability Ball Exercises

Exercise	OMNI RPE slide*
Squats (ball against a wall) Keep knees over ankles and squat to 90° angles at hips and knees	2
Calf raises (ball against a wall) Push up and into ball with foot as ankle extends, raising heel off ground	2
Seated knee extensions Extend leg to 180° to work entire quadriceps	2
Seated hip flexions Raise knee toward ceiling	2
Sit-ups Place middle back against ball, contract abdominal muscles	2
Back extensions Press upper abdominal muscles against ball, contract back extensors	2

* Have the client hold each position until the designated RPE slide is reached.

Adapted, by permission, from J. Schlacht, 2002, "Stability balls: An injury risk for older adults," *ACSM's Health and Fitness Journal* 6(4): 14-17.

Stretch Bands and Cords

Stretch-band (or -cord) exercises are a very popular component of core strength training programs for clients of all ages. Each repetition with a band employs the full range of joint motion, making strength outcomes functional and generalizable to a wide variety of fitness and sport activities. The bands are easy to manage, inexpensive, and can be used in group or individual workout sessions. The versatility of band exercises makes them especially suited for use in combination sets in cross-training workouts. Stretch-band workouts are guided by the same three primary RPE zones described earlier for general resistance training—endurance, hypertrophy, and strength. The RPE zone system for stretch-band exercise is used in this way:

- Select the band exercise to be used.
- Determine the OMNI RPE zone for the desired outcome: zone 3 for endurance, zone 6 for hypertrophy, or zone 9 for strength.

- Have the client use trial and error to select the band that has a "stretch feel" equal to the target OMNI zone.
- Begin the band exercise.
- Direct the client to perform repetitions until he reaches an OMNI RPE of 10, at which time the band exercise set is complete.
- Move the client to the next band exercise to work a different muscle group.
- Remember that one band repetition includes a complete contraction and relaxation of the target muscle group against the stretch band's resistance. The OMNI-RES should be in full view of the client at all times.

Mind–Body Programs

Two mind–body disciplines that are used widely in both group and individual workout sessions are Pilates and yoga. Both of these disciplines combine core stabilization with peripheral mobility training to enhance total body functioning. Quite often, stability ball and stretch-band exercises are incorporated into standard Pilates and yoga routines. Pilates exercises—and to a lesser extent some yoga poses—include sequentially performed dynamic movements. However, one choreographed component of both Pilates and yoga is a *held body* position requiring isometric muscle contractions. The duration for each held position can be guided by RPE slides on the OMNI scale as follows:

- Select the Pilates exercise or yoga pose.
- Once the client has assumed the "held" position, ask her for her RPE on the OMNI scale.
- Instruct the client to monitor her RPE while she holds the exercise position.
- When her RPE slides upward by 2 ratings, stop that exercise and move to the next exercise in the routine.
- The OMNI scale should be in full view of the client at all times.

RPE Preference Zone

Health-fitness clients who find training activities to be both pleasurable and effective are much more likely to adopt a regular exercise routine (Noble and Robertson 1996). One way to develop a conditioning program that meets these goals is to encourage clients to select the training intensities that they prefer. This allows clients to "set the pace" themselves (Buckworth and Dishman 2002). The RPE zone that corresponds to a client's preferred training pace is then determined. Most individuals' preferred RPE zone for aerobic conditioning lies between 10 and 15 on the Borg scale and 4 and 7 on the OMNI scale (table 6.10). Under most circumstances, an exercise pace that falls within the preferred RPE zone is considered a physiologically acceptable exercise training stimulus. The preferred zone is appropriate for such traditional conditioning modes as walking, running, and cycling and for comparatively new modes such as aerobic dance, water running, and high or low step aerobics. Whenever possible, prescribe an exercise training zone that uses the client's preferred level of exertion. A training intensity that improves aerobic fitness and is also perceptually acceptable promotes clients' compliance and helps them to establish a positive attitude toward physical activity (Buckworth and Dishman 2002).

Ensuring that the client's preferred intensity training zone elicits a physiological stimulus sufficient to improve physical fitness is important (Dishman 1994). For example, when clients are asked to select a jogging pace that is both comfortable and sustainable for a 30 min aerobic training session, they usually choose an intensity that falls between 50 and 75% of $\dot{V}O_2$max. This range corresponds to the training zone necessary to improve aerobic fitness (Dishman 1994). It is particularly important that the RPE zone that corresponds to a client's preferred training intensity is associated with the scale descriptors *very light, light,* and *somewhat hard* and that the client feels that the pace is tolerable over a prolonged training period.

Generally, the same preferred RPE training zone can be used by clients whose aerobic fitness varies from very low to very high levels (Dishman 1994). This is the case even though more-fit clients usually exercise at a higher intensity than less-fit clients. This ability to use a single preferred RPE training zone for all clients no matter what their self-selected training pace facilitates both individual and group exercise programming.

It is important to note that in cross-training programs, the preferred RPE zone may be influenced by the type of muscular action required (Noble and Robertson 1996). In general, RPE is lower for exercises involving concentric (shortening) as opposed to eccentric (lengthening) muscular actions. When a choice is available, urge clients to select exercises that minimize eccentric muscular actions.

Energy-Efficient RPE Zones

A preferred RPE training zone is especially useful when it can be linked to an **energy-efficient exercise intensity.** Normally, this is defined as an exercise intensity that requires the least energy and also has the lowest RPE. For example, when the resistance setting on a stationary cycle ergometer is constant, RPE is usually lowest for pedal rates between 60 and 80 rpm and highest when pedaling at either 40 or 120 rpm (Pandolf and Noble 1973; Robertson et al. 1995). Energy expenditure is also

Table 6.10 RPE Zones for Adult Clients According to Exercise Intensity and Type

Measurement	Preferred Intensity					
	Very light	Light	Moderate	Hard	Very hard	Maximal
% $\dot{V}O_2$max	<30	45	60	75	90	100
% HR_{max}	<60	70	80	85	95	100
RPE zone						
OMNI	1–2	4	6	7	9	10
Borg 6–20	7–9	11	14	15–16	18	19–20

From American College of Sports Medicine, 2000, *ACSM's guidelines for exercise testing and prescription* (Baltimore, MD: Lippincott, Williams, and Wilkins).

Exercise type			
	Target RPE zone		
Exercise machine*	OMNI	Borg 6–20	Feeling
Treadmill	1–9	7–18	Very light to very hard (+)
Cycle	1–9	7–18	Very light to very hard (+)
Arm and leg	4 6–7	11 14–16	Light (+) Moderate to hard (+)
Ski simulator	7 9	15–16 18	Hard (+) Very hard
Stepper	6 9	14 18	Moderate (+) Very hard
Rower	7 9	15–16 18	Hard (+) Very hard
Rider	6 7	14 15–16	Moderate (+) Hard (+)
Elliptical	4–7	11–16	Light to hard (+)
Slide board	2–9	8–18	Very light to very hard (+)
Rebounder	2–6	8–14	Very light to somewhat hard (+)
	Target RPE zone		
Free form	OMNI	Borg 6–20	Feeling
Walking	2–9	8–18	Very light to very hard (+)
Running	2–9	8–18	Very light to very hard (+)

(continued)

Table 6.10 *(continued)*

| Exercise machine* | Target RPE zone | | Feeling |
	OMNI	Borg 6–20	
Bicycling	2–9	8–18	Very light to very hard (+)
Swimming	2–9	8–18	Very light to very hard (+)
Dancing	6	14	Moderate (+)
Tennis	7	15–16	Hard (+)
Badminton	7	15–16	Hard (+)
Soccer	7–9	15–18	Hard to very hard
Calisthenics	4–7	11–16	Light to hard (+)
Hand weights, pump	6–7	14–16	Moderate to hard (+)
Aerobic dance	3–5	10–13	Light to moderate (+)
Water running	5	12–13	Light to moderate (+)

* Validated at CEHFR, University of Pittsburgh, by R. Robertson, F. Goss, K. Gallagher, J. Timmer, C. Dixon, T. Auble, N. Moyna, K. Lagally, and K. Wicker. % $\dot{V}O_2$max = percent of maximal oxygen consumption; % HR_{max} = percent of heart rate range; + = intensity within the preferred zone.

lowest at 60 to 80 rpm, making these rates energy efficient. When clients who are not competitive cyclists are allowed to select a pedal rate for a cycle training program, it normally falls within this energy-efficient and perceptually acceptable pace.

Table 6.10 summarizes RPE zones for a wide variety of weight-bearing and non-weight-bearing exercise training modes. By performing a given mode at an intensity that produces the indicated RPE zone, the client is exercising at a level that is often both perceptually preferable and energy efficient.

RPE Aquatic Zones

Aquatic exercise is a popular mode of training for health fitness as well as clinical rehabilitation (Greenberg, Dintiman, and Oakes 1998). The aerobic fitness benefits of aquatic exercise are similar to those obtained with land-based programs involving walking, running, and cycling. However, unlike these three modes, aquatic exercise simultaneously improves upper and lower body fitness. The non-weight-bearing effect of water buoyancy reduces stress on the hips, knees, and ankles and warm water provides a comfortable exercise setting, both of which promote regular participation. Because the body's weight is partially supported, overweight clients can exercise at comparatively higher target intensities than would normally be possible during land-based exercise. These comparatively higher aquatic training intensities contribute to both weight loss and improved aerobic fitness. In addition, because overweight clients are immersed in water, they report feeling less self-conscious about performing group exercise.

During aquatic exercise, intensity transitions are very important. Making smooth, gradual, and progressive intensity transitions is much easier when clients use sliding aquatic RPE zones. Table 6.11 shows how sliding zones can be used to guide an aquatic program. One advantage of these sliding RPE zones is that they apply to a wide range of aquatic training modes, many of which use supplemental devices such as webbed gloves, flexibility wands, and buoyancy belts or cuffs. Sliding zones enable clients to make smooth and manageable transitions from one exercise mode to the next within a single training session. The RPE zone that is typically produced during the stimulus phase of an aquatic program is equivalent to the anaerobic threshold for most modes of water training—5 to 7 on the OMNI scale and 12 to 14 on the Borg 6–20 Scale (Frangolius and Rhodes 1995). A practical note regarding the optimal pace of various aquatic activities may be helpful to the practitioner. In general, when the aquatic exercise intensity is low, the optimal and perceptually desirable pace for stepping, striding, or sculling is slower (approximately 40 rpm). When the intensity is high, faster rates (approximately 70 rpm) may be perceived by the client as more desirable (Robertson et al. 1995).

Table 6.11 Sliding RPE Aquatic Training Zones

| Component | RPE zone | | Duration (min) | Feeling |
	OMNI	Borg 6–20		
Warm-up				
Thermal warm-up	1	7	3–5	Easy
Prestretch	1	7	3–5	Easy
Cardiorespiratory	3	10	3–5	Somewhat easy
Stimulus (aerobic)*				
Continuous, interval, or circuit	5–7	12–16	20–60	Somewhat hard to hard
Cool-down				
Cardiorespiratory	3	10	3–5	Somewhat easy
Final stretch	1	7	5–10	Easy

* Dancing, deepwater running, stepping, striding, sculling, walking, or jogging. Many of these activities can be performed using wrist or ankle weights, hand paddles, fins, and pull buoys.

From Aquatic Exercise Association, 1995, *Aquatic fitness professional manual* (Nokomis, FL: Aquatic Exercise Association).

The training zones listed in table 6.11 use both the OMNI and Borg scales. Practitioners may also want to use Morgan's special 7-category scale for aquatic exercise measurements, which is included in appendix A (Morgan 2001). A shortened version of the Borg 6–20 Scale, is ideally suited for lap-swimming programs. Because the scale employs only 7 categories, it is easy for a client to use it to guide lap swim times by RPE.

Water Running Using RPE Zones

Deepwater running is an excellent non-weight-bearing activity for aerobic conditioning, weight loss, and injury rehabilitation. Clients with adequate aquatic ability who are comfortable in the water can perform a water-running routine with or without small arm or shoulder flotation devices. The most effective method for guiding the intensity of a water-running program is the sliding RPE zone system. Note that for most forms of water-immersion exercise, the use of RPE zones to guide intensity is preferable to target HR. This is because the HR is typically lower during water- than during land-based exercise, making the transferring of exercise prescriptions between track and pool difficult (Svedenhag and Seger 1992). In addition, HR is difficult to measure during water running and varies markedly as body position and depth of immersion change, making it impractical to control water-exercise intensity using a target HR (Wilder and Brennan 1993). The sliding RPE zone system automatically adjusts exercise intensity to accommodate the water medium and, most importantly, is much easier to use than the troublesome pulse-counting method.

The following is a 6-day water-running program that employs the sliding RPE zone system (table 6.12). This program is adapted from the original format developed by Bushman and colleagues (1997). For convenience, the RPE intensity zones have been converted to the OMNI walking and running picture scale.

RPE Zones for Weight Loss

Significant health benefits can be achieved by increasing energy expenditure with low- to moderate-intensity activities (Moyna et al. 2001). Expending 200 kcal per day through physical activity can result in substantial loss of excess body weight when the conditioning program is combined with a modest reduction in daily caloric intake of approximately 400 kcal. It is estimated that the caloric deficit resulting from this combination of diet and exercise causes a loss of 1 lb of body fat every 6 days. Although the total exercise time required to achieve the target level of energy expenditure varies somewhat between women and men, it seldom exceeds 30 min. Table 6.13 lists the length of time required to

Table 6.12 Deepwater Running Training Program

Day	Workout (min:s)	Total session time (min)	Cycles/ min	OMNI RPE zone*	Feeling*
Mon	5 × 2:00	28	60–75	7	Somewhat hard
	8 × 1:00		75–85	9	Hard
	5 × 2:00		60–75	7	Somewhat hard
Tues	7:00	31	60–75	7	Somewhat hard
	6:00		60–75	7	Somewhat hard
	5:00		60–75	7	Somewhat hard
	4:00		60–75	7	Somewhat hard
	3:00		75–85	9	Hard
	2:00		75–85	9	Hard
	4 × 1:00		75>85	9–10	Hard to extremely hard
Wed	45:00	45	75>85	9–10	Hard to extremely hard
Thurs	3 × 3:00 and	36	50–75	4–7	Somewhat easy to somewhat hard
	3 × 1:00; repeat × 3		75>85	9–10	Hard to extremely hard
Fri	8:00	35	50–60	4	Somewhat easy
	7:00		60–75	7	Somewhat hard
	6:00		60–75	7	Somewhat hard
	5:00		60–75	7	Somewhat hard
	4:00		60–75	7	Somewhat hard
	3:00		60–85	7–9	Somewhat hard to hard
	2 × 1:00		75>85	9–10	Hard to extremely hard
Sat or Sun (optional)	45:00	45	50–75	4–7	Somewhat easy to somewhat hard

Equivalent land exercise with RPE zone and workout: 1, slow walk (>21 min/mile); 4, medium-pace walk (15–20 min/mile); 7, fast walk or jog (<15 min/mile); 8–9, run (5–10 min/mile); 10, very hard run (<5 min/mile)

From B.A. Bushman et al., 2000, "Effect of 4 wk of deep water run training on running performance," *Medicine and Science in Sports and Exercise* 29(5): 694-699; and IDEA Health & Fitness Association, May 2000. Reproduced with permission of IDEA Health & Fitness Association, (800) 999-IDEA, www.IDEAfit.com

* Added by R.J. Robertson as a supplement to this table.

expend 200 kcal of energy for six weight-bearing and non-weight-bearing exercise machines used at RPEs of 11, 13, and 15 on the Borg 6–20 Scale and 4, 5, and 6 on the OMNI scale. For example, this table can be used to prescribe an exercise weight-loss program for a female client where the target energy expenditure is 200 kcal per training session. For a treadmill exercise program, the target energy expenditure can be achieved in 21 min at an RPE of 11 and 17 min at an RPE of 15. Several different modes can be incorporated into the training program by having clients slide their RPE training zone between, for example, 11 and 15. That is, a male client could exercise for approximately 14 min on a treadmill at a target RPE of 11 or for 13 min on a stair stepper at a target of 15. For both modes, the total energy expenditure during the training session is approximately 200 kcal, so combining the two in an exercise session by using a sliding RPE zone adds flexibility and variety to the training program. Clients generally view such programs as being more pleasurable, which ultimately increases program adherence—a critical requirement for long-term weight loss.

Table 6.13 Time Required to Expend 200 kcal During Aerobic Exercise at Three Target RPEs

Exercise machine	Gender	Time (min) at target RPE*		
		RPE 11/4	RPE 13/5	RPE 15/7
Treadmill	Male	14:00	12:15	11:15
	Female	21:00	18:00	17:00
Ski simulator	Male	15:30	13:00	12:00
	Female	22:15	20:30	18:45
Stair stepper	Male	19:15	15:15	13:00
	Female	35:00	23:00	20:15
Rower	Male	20:45	16:15	13:30
	Female	24:00	20:15	17:30
Rider	Male	23:45	21:00	19:00
	Female	35:00	30:15	27:15
Cycle	Male	27:45	21:45	16:00
	Female	43:45	30:15	23:15

* Borg 6–20/OMNI scales.

Adapted, by permission, from N.M. Moyna et al., 2001, "Intermodal comparison of energy expenditure at exercise intensities corresponding to the perceptual preference range," *Medicine and Science in Sports and Exercise* 33(8): 1404-1410.

RPE Weight-Loss Zones for the Home Exerciser

Significant weight loss can also be realized by the client who participates in 5 to 10 min blocks of aerobic activities for up to 30 min per day. The Physical Activity and Weight Management Center at the University of Pittsburgh recommends that these intermittent bouts of physical activity be performed at intensities that alternate between a moderate RPE zone of 10 to 12 on the Borg scale and 3 to 4 on the OMNI scale and a vigorous RPE zone of 13 to 15 (Borg) and 5 to 7 (OMNI). Clients who accumulate up to 200 min per week of exercise at these RPE weight-loss zones can expect to lose as much as 17 lb over a 12-month period (Jakicic et al. 2003). To help clients guide their exercise intensity, give them a playing-card-size Borg or OMNI RPE scale with the target RPE zones highlighted for easy reference.

Flexibility

Flexibility is the ability to move a body part through its full range of motion. To improve flexibility, stretching exercises should be performed during the warm-up and cool-down phases of each workout session. Two types of stretching exercises generally are recommended for health-fitness clients: slow static stretching and proprioceptive neuromuscular facilitation, or PNF (Hoffman 2002). The range of motion used for these stretching techniques can be controlled by an upper-limit RPE that is specifically linked to the muscle groups being stretched.

Static stretching involves slowly stretching a muscle to its farthest point at an upper-limit RPE of 7 or 8 on the OMNI scale. This position is held for 10 to 30 s. Static stretching is particularly useful for clients who are just beginning an exercise program. Its slow, prolonged movement minimizes pain and injury. In addition, the movements are easy to learn.

PNF stretching usually is performed with a partner, such as the practitioner or another client. It uses a contract–relax technique. The muscle is first stretched by the partner, who provides slow movement force on the limb in the direction of the stretch. When the muscle has been stretched to the point of slight discomfort (at an RPE of approximately 3 on the OMNI scale), the client isometrically contracts the muscle for 5 to 15 s. The muscle is then briefly relaxed, with the limb remaining in the stretched position. The partner then applies slow pressure to further stretch the muscle, stopping when the upper-limit RPE of 8 or 9 (OMNI) is reached. This second assisted stretch requires 10 to 30 s to complete. Three to four repetitions should be performed for each stretching exercise. The PNF stretching technique should be used by more-advanced clients. Although it is very effective, the degree of pain or discomfort can be high and the procedure requires skill, practice, and a partner.

SUMMARY

This chapter examines RPE training zones for both aerobic and resistance exercise, focusing primarily on the health-fitness client. A target RPE training zone (a) improves aerobic fitness, reduces excess body weight, and increases muscular strength and (b) is appropriate for clients who have certain characteristics in common, such as age, fitness level, gender, and training experience. Practitioners should use the RPE conversion chart (figure 6.1) to help them transfer training zones from one scale to another. The application of the RPE training zone that is equivalent to the anaerobic threshold is explained. The use of sliding RPE zones in single and combination land and aquatic exercises is explained and workout programs are provided. Finally, upper-limit RPEs used to guide stretching and flexibility exercises are described.

C·H·A·P·T·E·R

7

Rating of Perceived Exertion Zones for Special Clients and Conditions

CASE STUDY

Client Characteristics

The client is a 24-year-old, clinically normal female who participates in aerobic exercise (jogging, cycling, and swimming) three to five times per week. She is 6 weeks into her second pregnancy. The client exercised regularly throughout her first pregnancy without maternal or fetal complications.

Exercise Need

The client had a very satisfactory exercise experience during her first pregnancy. Both maternal and fetal weight gain were optimal during pregnancy, and the postpartum return to prepregnancy body weight and fitness level was rapid. Therefore, the client has requested that the practitioner design an exercise program that will be appropriate for her fitness needs throughout her second pregnancy.

Action Plan

With physician consultation, the practitioner developed a 3-days-per-week exercise program that prescribed light- to moderate-intensity weight-bearing (walking, jogging, or stepping) activities in the early and middle phases of pregnancy and non-weight-bearing (cycling or swimming) activities in the pregnancy's final phase. The client's HR during the stimulus portion of each workout session did not exceed 144 beats per minute (bpm). This intensity generally corresponded to an RPE conditioning zone of 4 to 5 on the OMNI scale or 11 to 13 on the Borg 6–20 Scale.

An increasing number of special exercise programs offered by fitness clubs and community recreation centers focus on the health-fitness needs of female and older clients. The unique exercise needs of both of these groups often require specially developed training and physical activity programs. In addition, many children and young adults are not sufficiently active and have low levels of physical fitness and excess body fat. One way to meet the exercise needs of children and adolescents is to focus the physical education curriculum on activities that improve fitness, promote a healthy lifestyle, and encourage regular participation. Physical activity and sport conditioning programs specifically designed for each of these special groups can use RPE zones as a simple and effective method for guiding exercise intensity. The first part of this chapter examines how RPE zones are incorporated into special conditioning programs for women, older clients, and children.

RATING OF PERCEIVED EXERTION ZONES FOR WOMEN, OLDER CLIENTS, AND CHILDREN

Women, older clients, and children all present unique challenges to the health-fitness professional. To meet these challenges, special physical activity programs can be developed that employ RPE zones for both aerobic and resistance exercise training.

Exercise Programs for Women

In general, women and men respond to health-fitness programs similarly, realizing the same benefits with regular participation (Greenberg, Dintiman, and Oakes 1998; Luxbacher, Bonci, and King 2002). Therefore, the RPE training zone systems described in chapter 6 are appropriate for both women and men. Many health-fitness clubs have conditioning programs specifically tailored for the needs of their women clients. The benefits of these programs include reduced body fat, increased lean tissue, improved blood lipid levels, lower blood pressure, less emotional stress, and lower risk of cardiorespiratory, metabolic, and carcinogenic diseases. Women of all ages can undertake appropriately prescribed aerobic and resistance exercise training programs. Those who do can improve their aerobic fitness, cardiorespiratory endurance, muscular strength, and sport performance.

A conditioning program specially designed for women should have aerobic exercise, resistance exercise, and flexibility components. Aerobic exercise should be performed 5 to 7 times per week, with workout times of at least 12 to 30 min per session. Beginning exercisers may need to pause briefly (for approximately 30 s) to rest every 4 min. The goal is to perform longer and longer periods of exercise before pausing for a rest. Aerobic conditioning activities that are often preferred by women include

- aerobics,
- ice skating,
- cross-country skiing,
- step aerobics,
- swimming,
- tennis,
- running,
- exercise walking, and
- roller or in-line skating.

A sliding RPE training zone can guide the intensity of each of these aerobic activities. (See chapter 6 for an explanation of the sliding zone system.) In general, the warm-up and cool-down zones should be performed at an OMNI scale RPE of 1 to 3 (7 to 10 on the Borg scale). The exercise itself should be performed at intensity zones that slide up and down from RPE 4 to 7 on the OMNI scale (11 to 15 on the Borg scale).

The resistance exercise component of the training program can also be guided by the sliding RPE zone system. Most women prefer resistance exercise programs that emphasize development and maintenance of local muscular endurance and give some attention to muscular strength. Muscle hypertrophy usually is not a training objective for female clients. To improve muscular endurance, use the RPE 3 endurance zone shown on the specially formatted OMNI-RES in figure 6.4. When training in the endurance RPE zone, the resistance or weight will be somewhat lighter than that used for muscular strength and the client will likely do 10 to 15 repetitions for each set. Two to three sets can be performed for each exercise, with a 30 to 60 s rest between sets. If developing muscular strength is important, use the RPE 9 (strength) zone on the scale. Follow the training procedure described in chapter 6. A very good description of resistance exercises specifically designed for women can be found in *Total Fitness for Women* (Luxbacher, Bonci, and King 2002).

Table 7.1 shows a total body resistance exercise training program for women. The amount of weight to be lifted and number of repetitions to be performed are guided by OMNI RPE training zones as follows:

- For each exercise, select the desired RPE zone, such as zone 3 for endurance training.
- Have the client select, by trial and error, a training weight that produces the target RPE zone.

- Begin the exercise.
- Instruct the client to continue with the exercise repetitions until she reaches an OMNI RPE of 10, when the set is complete.
- Move to the next resistance exercise.

Remember that one exercise repetition includes a complete contraction and relaxation of the target muscle group against the resistance. The OMNI scale should be in full view of the client at all times.

Flexibility is a very important dimension of a total fitness program for women. Flexibility exercises should be included in both the warm-up and cool-down phases of each workout session. A static stretching procedure guided by an upper-limit RPE (see chapter 6) should be used. An excellent list of exercises for total body flexibility is presented in Luxbacher, Bonci, and King (2002).

Table 7.1   Strength Training Program for Women Using OMNI-RES RPE Zones

Exercise	Muscle group	RPE zone	Number of sets*	Number of times per week
Leg press	Quadriceps	3	1–2	2–3
Chest press	Pectoralis	3	1–2	2–3
Triceps extension	Triceps	3	1–2	2–3
Latissimus pull-down	Latissimus dorsi	3	1–2	2–3
Hamstring curl	Hamstring	3	1–2	2–3
Biceps curl	Biceps	3	1–2	2–3
Calf raise	Gastrocnemius	3	1–2	2–3
Lateral raise	Medial deltoid	3	1–2	2–3
Abdominal curl	Rectus abdominis	3	1–2	2–3
Lower back extension	Erector spinae	3	1–2	2–3

* Rest for 30 to 60 s between sets and 60 s between exercises.

Adapted from IDEA Personal Trainer, 2002, "Women's strength program." Reproduced with permission of IDEA Health & Fitness Association, (800) 999-IDEA, www.IDEAfit.com.

One area of health-fitness programming that is of growing interest to women is physical activity participation during pregnancy (Greenberg, Dintiman, and Oakes 1998). The decision to exercise during pregnancy should always receive physician approval. Most physicians advise patients who were physically active prior to pregnancy to remain active, at least during the early to middle phases of pregnancy. In general, women who exercise during pregnancy have fewer complications, weigh less at delivery, and gain less weight. A safe and effective exercise program for the pregnant client employs light to moderate aerobic activities over the first 3 to 6 months with gradual tapering or in some cases cessation over the 3 months prior to delivery. Most women find non-weight-bearing activities such as cycling and swimming to be preferable, especially during the final trimester. It is recommended that aerobic exercise be performed at least three times per week. Normally, the mother's heart rate should not exceed 144 bpm during the stimulus period. For younger women, a heart rate of 144 bpm generally corresponds to an RPE training zone of 4 to 5 on the OMNI scale (11 to 13 on the Borg scale). Warm-up and cool-down activities should be performed in OMNI RPE zone 1 to 2 (7 to 9 on the Borg scale). In some cases, pregnant clients may find that they prefer regulating their exercise intensity with a target RPE zone rather than an absolute HR cutoff point. This is because exercise HR can vary from day to day in response to maternal and fetal changes that become more pronounced as the pregnancy progresses. In contrast, RPE training zones are stable because they automatically adjust to accommodate bodily changes throughout pregnancy.

Aquatic Exercise for the Pregnant Client

Aquatic exercise is ideally suited for pregnant clients. If the client has followed an exercise program before the pregnancy, the objective of aquatic training should be to maintain current fitness levels rather than to increase the training dosage. The water's buoyancy reduces stress on bones and joints, facilitates balance and stability, promotes heat dissipation, and provides a comfortable and secure exercise environment (Aquatic Exercise Association 1995). In many cases, water programs allow clients to continue fitness activities throughout much of their pregnancy. In contrast, land-based exercise can become uncomfortable and in some instances unsafe during the latter stages of pregnancy. For pregnant clients, it is important that the change in aquatic exercise intensity from warm-up to stimulus to cool-down be smooth and gradual. The sliding RPE zone system is ideally suited to guiding intensity transitions between phases of the aquatic program.

Physical Activity and Osteoporosis

Physicians recommend regular exercise as a way to slow bone loss and help delay or prevent osteoporosis. Osteoporosis, which is caused by a decrease in bone density, occurs in both women and men as they age. However, because it is twice as prevalent in women, it is often viewed as a special health concern that should be addressed through physical activity and dietary programs developed specifically for women.

Exercise places an adaptive stress on bones, causing them to strengthen (Greenberg, Dintiman, and Oakes 1998). That is, as the muscles contract during exercise they pull on the bones to which they are attached. This stress causes the bones to increase their density and become stronger. Bone density can be increased with such activities as walking, jogging, aerobic stepping, dancing, calisthenics, and aquatic exercises. The intensity of these aerobic activities can be guided by the RPE zones given in chapters 5 and 6. Aerobic exercise programs should be approximately 30 min in duration and take place two to three times per week. Practitioners should remember that the sliding RPE zone system is especially useful for clients with osteoporosis. By sliding from comparatively low to moderate intensities, the client can gradually adapt to the exercise stimulus without placing undue stress on weak bones. Table 7.2 presents an aerobic training program for clients who have osteoporosis. Each activity is guided by an

Table 7.2   Combination Aerobic Training Program for the Client with Osteoporosis

Set	Exercise	OMNI RPE zone*	Feeling*
1	Pedal on stationary bike for 4 min	1–2	Easy
2	Use recumbent stepper for 4 min	4–6	Somewhat easy to somewhat hard
3	Step on and off a 6 in. step or platform for 2 min	4–6	Somewhat easy to somewhat hard
4	Use upper arm ergometer for 3 min	4–6	Somewhat easy to somewhat hard
5	Sit in wheeled office chair and wheel across floor for 2 min	1–2	Easy
	Total exercise time: 15 min		

Rest for 1 min between sets (or longer if necessary).

From J.H. Rimmer, 1999, "Programming for clients with osteoporosis," IDEA Personal Trainer. Reproduced with permission of IDEA Health & Fitness Association, (800) 999-IDEA, www.IDEAfit.com.

* Added by R.J. Robertson as a supplement to this table.

RPE zone. However, it is important that clients understand that no matter what RPE zone is to be reached, unusual pain in arthritic areas indicates that the exercise level is inappropriate. The intensity or duration of the activity should be reduced until the painful symptoms disappear or at least fall within the client's normal, tolerable range.

Resistance exercise training is also an excellent method for preventing bone loss. In general, women lose about 1% of their muscle mass each year after the age of 30. As a result, muscles become weaker and less defined. In conjunction with this loss of muscle mass, bones become weaker and more brittle. Strength training helps to minimize the negative effects of aging on muscle and bone and promotes a more functional lifestyle. Women who wish to offset the deleterious effect of osteoporosis should use RPE zone 3 (endurance) on the OMNI-RES. The endurance zone employs light weights and requires 10 to 15 repetitions per set. A resistance exercise program should include 8 to 10 exercises that together cover all the major muscle groups of the body. Training should be performed at least twice per week.

Osteoporosis progresses as clients age. Therefore, lifelong participation in weight-bearing exercise programs is recommended to delay or prevent bone loss. The use of RPE zones to guide exercise intensity is particularly helpful in this regard. As clients age, they will automatically adjust their exercise intensity to produce the target RPE zone recommended for good bone health.

RPE Conditioning Zones for the Cancer Survivor

Regular exercise is increasingly being prescribed as an important part of a recovery program for cancer patients. This is especially true of breast cancer patients who have undergone surgery or chemotherapy. Table 7.3 presents a combination aerobic, resistance, and mind–body exercise program for cancer survivors that uses RPE zones.

Does Gender Influence RPE?

Practitioners often wonder whether RPE differs between female and male clients. This important question is particularly pertinent in developing exercise programs for groups that include females and males. The answer is both *no* and *yes* (Robertson et al. 2000b). In general, when females and males perform aerobic exercise at the same % $\dot{V}O_2$max, their RPEs will not differ. However, when females and males perform at the same absolute exercise intensity (or training pace), the RPE will be higher for the client who has the lower aerobic fitness level. Given equal levels of fitness, females usually have a lower $\dot{V}O_2$max than male clients. Therefore, the female client's RPE is higher than the male's when the two perform at the same absolute aerobic exercise intensity.

Table 7.3   Exercise Program for the Cancer Survivor

Modality	Time (min)	OMNI RPE zone	Comments
Aerobic exercise training (machines, walking, stepping, water aerobics)	5–15	2–4	Basic warm-up
Strength training (machine circuit, dumbbell, stretch-band exercises)	25–40	3 (muscular endurance)	Work major muscle groups for 2–3 sets, 1–2 sets for smaller muscle groups; 8–12 exercises per workout
Stretching and range of motion	10–15	Upper-limit RPE of 7–8	Use individual or partner stretches for areas affected by surgery and major joints (shoulders, hips, low back) for general improvement
Mind–body training	10	2–4 RPE slides	Breathing, relaxation exercises, yoga, movement therapy

Adapted, by permission, from E. Durak, 2001, "The use of exercise in the cancer recovery process," *ACSM's Health and Fitness Journal* 5(1): 6-10.

Similarities and differences in the RPE responses of women and men are also observed during resistance exercise. When female and male clients lift the same % 1RM, their RPE-AMs and RPE-Os are the same (Pincivero et al. 2001; Robertson et al. 2003). However, when female and male clients lift the same absolute weight and, as is usually the case, the male is stronger, the female's RPE is higher (O'Connor, Poudevigne, and Pasley 2002). This holds true for both isotonic and isometric muscular actions.

It is also asked whether female and male children differ in their perceptions of physical exertion. The answer is *no* when comparisons are made at a given $\%\dot{V}O_2max$. However, there is no set pattern of similarities and differences in RPE between prepubescent female and male children when comparisons are made at an absolute exercise intensity. The child—regardless of gender—who has the highest $\dot{V}O_2max$ has the lowest RPE when performing at a given exercise intensity or training pace. It was shown that RPEs derived from the children's OMNI scale did not differ between female and male children (6 to 12 years old) performing cycle ergometer and treadmill exercise (Robertson et al. 2000a; Utter et al. 2002) or when self-regulating intermittent cycle ergometer exercise intensity according to a target training RPE.

Therefore, some differences in RPE between females and males can exist during both aerobic and resistance exercise. In general, if an exercise group includes both female and male clients, prescribe an intensity that is based on the same percentage of maximum for all clients. Doing so will allow the group to remain reasonably cohesive as the program progresses, which will promote camaraderie and facilitate supervision. However, it should be remembered that if female and male clients are exercising at the same target RPE, the client with the highest level of fitness—regardless of their gender—will perform at the faster pace or lift the heavier weight.

RPE Zones for the Older Client

It is estimated that only 30 percent of Americans 65 years of age and older exercise on a regular basis (Greenberg, Dintiman, and Oakes 1998; U.S. Department of Health and Human Services 1996). Exercise should be an important part of the daily routine of older women and men. Older clients who participate regularly in some form of exercise usually have higher levels of physical and psychological functioning. They are comparatively less dependent on others, more pain free, less prone to falling, and more varied in their daily activities.

For older clients, participating in regular physical exercise reduces and in some cases prevents functional declines associated with aging (ACSM 1998). Older individuals can benefit from both aerobic and strength training programs. Aerobic training helps maintain and improve cardiovascular function, reduces the risk of developing heart disease and diabetes, improves health status, and increases life expectancy. Resistance training helps to offset the loss of muscle mass and strength typically associated with normal aging. Stretching, flexibility, and balance activities help to prevent falls. In addition, regular physical activity preserves cognitive function, relieves depression, and improves perceptions of self-worth and self-sufficiency. The combined effect of these training adaptations is to improve both functional fitness and the quality of life in older clients.

The goal of an exercise program for the older client should be to improve functional performance, functional capacity, or a combination of the two (Bonder and Wagner 2001). These functionally based exercise programs stimulate the cardiovascular, neuromuscular, and skeletal systems. In general, the exercise intensity is lower when the goal is to improve functional performance and higher when the goal is to improve functional capacity. Light- to moderate-intensity lifestyle activities optimize health. Moderate- or high-intensity exercise may be required to improve aerobic fitness and reduce disease risk (ACSM 1998). Whether the conditioning goal is to increase functional performance or functional capacity, RPE training zones can play a key role in helping older clients regulate the intensity of their conditioning program by making daily exercise enjoyable and easy to perform and eliminating troublesome and sometimes expensive electronic HR-monitoring devices. Because RPE zones automatically guide exercise intensity, the client's attention does not need to focus on the exercise task, which promotes interaction between clients during activity sessions.

Table 7.4 presents RPE training zones for aerobic activities according to the fitness status of the older client. A general list of conditioning activities classified by training outcome for older clients appears in table 7.5. Note

Table 7.4   RPE Aerobic Training Zones for the Older Client

Estimated functional capacity	RPE zone		Intensity (%)	Duration (min)	Frequency
	OMNI	Borg 6–20			
High	7–8	15–17	70–85	20–40	3–5/week
Intermediate	5–7	12–16	60–75	20–30	1–2/day
Low	3–7	10–16	40–75	10–30	2–3/day

From B.R. Bonder and M.B. Wagner, 2001, *Functional performance in older adults*, 2nd ed. (Philadelphia, PA: F.A. Davis).

Table 7.5   Recommended Physical Activities for the Older Client

Training outcome	Activities
Aerobic	Walking, jogging, machine stair stepping, cycling, swimming, rowing, aerobic dancing, aquatic exercising, or swimming
Recreational	Gardening or yard work; golfing; ballroom dancing; playing shuffleboard, boccie, croquet, horseshoes
Resistance	Exercises using all major upper and lower body muscle groups (preferably with isotonic contractions)
Flexibility	Extent and static stretching, calisthenics, walking, aerobic dancing
Fall prevention	Balance training, resistance exercises, walking, weight-transfer activities

Perform weight-bearing and non-weight-bearing activities on alternating days.

Compiled from ACSM 2000; Bonder and Wagner 2001; Greenberg, Dintiman, and Oakes 1998.

that these conditioning activities are suggested for older clients whose physicians have determined them to be clinically capable of performing at the requisite effort levels. Chapter 9 describes RPE training zones to be used by older adults who have clinically limited exertional tolerance.

Resistance Exercise for the Older Client

Table 7.6 sets out a 6-week progressive resistance exercise program for older clients. Each exercise uses RPE zone 3 (endurance) on the OMNI-RES. The RPE zone system is used as follows:

- Select the exercise that is appropriate for the muscle group(s) to be trained.
- By trial and error, select a training weight that produces the target RPE zone of 3.
- Begin the exercise.
- Have the client continue with the exercise repetitions until he reaches an OMNI RPE of 10, when the set is complete.
- Move to the next resistance exercise.

Remember that one exercise repetition includes a complete contraction and relaxation of the target muscle group against the resistance. The OMNI scale should be in full view of the client at all times.

Exercise for the Arthritic Client

Some older clients report that the discomfort of arthritis reduces their ability to exercise on a regular basis. However, exercise programs for arthritic clients that employ RPE zones make the activities safe, comfortable, and effective. The goal of the exercise program is to help the client regain or maintain joint range of motion and functional movement skills (Greenberg, Dintiman, and Oakes 1998). Consultation with the client's physician is important in identifying the appropriate exercise dosage. The upper level of the RPE training zone should be determined individually for each client and set just below the intensity that produces pain. It may be useful to measure the client's arthritic pain prior to starting an exercise session and then determine the increase in pain from this baseline measure during the actual session. Chapter 9 discusses a pain scale that can easily be used during exercise. Once it has been determined, the client exercises at her *symptom-free* RPE training zone during the stimulus portion of the activity program. The 2 h pain rule should be observed: If the client experiences pain or soreness for more than 2 h after a workout, the intensity or duration of the next session should be reduced. Exercise programs for arthritic clients should have a longer warm-up, a lower

Table 7.6 Resistance Exercise for the Older Client Using RPE Zones

Week	Exercise	OMNI RPE zone	Number of sets	Objective
	Start			
1–2	Leg press	3	1–2	To develop foundational muscular endurance using 4 basic exercises that involve multijoint, linear, pushing, and pulling movements
	Chest press	3	1–2	
	Compound row	3	1–2	
	Abdominal curl	3	1–2	
	Add			
3	Overhead press	3	1–2	To continue foundational progression using same movements as in weeks 1–2
	Lower back extension	3	1–2	
	Neck flexion and extension	3	1–2	
	Add			
4	Leg extension	3	1–2	To emphasize single-joint flexion and extension movements
	Leg curl	3	1–2	
	Arm extension (triceps)	3	1–2	
	Arm curl (biceps)	3	1–2	
	Add			
5	Chest fly	3	1–2	To emphasize single- and multijoint rotational movements
	Rotary torso	3	1–2	
	Lateral raise	3	1–2	
	Super pull-over	3	1–2	
6	All 15 exercises	3	1–2	To exhibit sufficient foundational muscular strength and endurance for a 15-exercise program involving major muscle groups

From B.A. Pruitt, 2003, "Exercise progressions for seniors," IDEA Personal Trainer. Adapted with permission of IDEA Health & Fitness Association, (800) 999-IDEA, www.IDEAfit.com.

intensity, and fewer repetitions than programs for nonarthritic clients. The activities should involve all of the major muscle groups. It is also important to include in each exercise session fine motor skills involving the fingers, wrists, ankles, and feet. Submerging the affected limb in warm water during the exercise bout helps to reduce pain and improves the range of motion.

RPE Zones for Children

Children often prefer physical activities that are intermittent and free-form. To promote a healthy lifestyle, at least a portion of these activities should involve dynamic muscular contractions and stress the aerobic energy system. Children can self-regulate their exercise intensity with sliding RPE training zones. Ask the child to produce a target RPE that slides from low to moderate to high intensity (Robertson et al. 2000a; Robertson et al. 2002). Use the Children's OMNI Scale of Perceived Exertion to guide these sliding target RPE zones (see chapter 2 for an explanation of OMNI scale development). The children's OMNI scale can be used for a wide range of aerobic and resistance exercises by selecting the format that has pictorial cues depicting the activity to be performed. In addition, children can use the OMNI scale to differentially rate their RPE-O as well as the RPEs for the legs, arms, and chest. Normally, all four ratings can be made within a 30 s period.

Anaerobic Threshold Zone for Children

The most useful RPE training zone for children employs the anaerobic threshold as the prescriptive reference point. As noted in chapter 6, the anaerobic threshold can be used to prescribe both continuous and intermittent exercise training programs for aerobic conditioning and weight loss. Identifying the stable RPE zone that corresponds to the anaerobic threshold helps children to guide the intensity of their exercise program. For children, the OMNI scale RPE zones that correspond to the anaerobic threshold are 6 for the overall body, 4.5 for the chest, and 7 for the legs. In general, these OMNI scale RPE zones can be used with 8- to 14-year-old boys and girls whose aerobic fitness levels vary from low to high. On the Borg 6–20 Scale, the RPE zone that corresponds to the anaerobic threshold is 11 to 14 in 8- to 12-year-old children (Mahon et al. 1998).

Children's play often involves short periods (1 to 3 min) of moderate- to high-intensity exercise interspersed with brief periods of low-intensity exercise or rest. The sliding RPE zone system is ideally suited to helping children guide the intensity of their free-form play. The exercise format presented in table 7.7 should be used for such aerobic exercise modes as walking, running, cycling, stepping, jumping, and sliding.

Table 7.7   RPE Sliding Zone Training System for Children

	Warm-up	Stimulus***	Cool-down
Time (min)*	1–2	3–3–3–3–3	1–2
OMNI RPE zone**	2	3–7–3–7–3	2

* Additional alternating time periods can be included as the child's aerobic fitness improves.

** A sliding zone that uses a target RPE of 8 or 9 can be added as the child's aerobic fitness improves.

*** The same or a number of different aerobic exercise modes can be used for the stimulus period.

This sliding zone system uses as the prescriptive reference the target RPE on the OMNI scale that is equivalent to the anaerobic threshold. The exercise then slides between an intensity that is less than the anaerobic threshold and one that is slightly greater than it. A sliding zone that uses a target RPE substantially higher than the anaerobic threshold can be employed as aerobic fitness improves. It is important that the majority of the exercise stimulus period be spent at or above the target zone equivalent to the anaerobic threshold. Exercise performed at the child's anaerobic threshold provides a training stimulus to increase aerobic fitness and promote cardiovascular health.

Because sliding RPE zones resemble children's own play activities, they view the training programs as fun. Sliding between the low, moderate, and high target zones adds flexibility to the program, and children often treat the attainment of different RPEs for different types of exercise as a game. In general, sliding RPE zones encourage children to sustain their exercise for long enough to provide a training stimulus. Because the activity is fun and challenging, children look forward to the exercise sessions, which promotes long-term program adherence. They like the OMNI scale's pictures, too.

RPE Age Threshold

The pioneering research involving RPE measurements in children performed by Oded Bar-Or, M.D., was first reported in 1977 (Bar-Or 1977). One of his earliest studies was undertaken to determine the **RPE age threshold**—the youngest age at which a child could effectively use an RPE scale. Since these early investigations, it has been determined that children as young as 6 years of age can effectively use the OMNI and Borg 6–20 scales and CERT to rate their perceived exertion. This RPE age threshold is appropriate for exercise testing, self-regulating training intensity, and tracking training progress.

ENVIRONMENTAL INFLUENCES ON RATING OF PERCEIVED EXERTION ZONES

RPE is a valuable tool for monitoring the additional strain imposed by hot and cold air temperatures and high elevation. Many health-fitness clients like to exercise outdoors. One of the most common environmental problems encountered when exercising outdoors is a high air temperature, often accompanied by high humidity levels. Perceived exertion during dynamic exercise becomes more intense as the air temperature gets hotter. This suggests that RPE is influenced by physiological processes that regulate body temperature during exercise (Pandolf 2001). Therefore, RPE serves two very important functions in training programs for clients who exercise where the ambient temperature is occasionally very high and potentially dangerous. First, the RPE response can be used as a marker of the client's level of acclimatization to high outdoor temperatures, and second, the RPE response can be used to make day-to-day adjustments in training intensity to prevent heat strain without compromising the overload training stimulus.

High Temperature and RPE

At a normal training pace, RPE is higher in high air temperatures, especially when the relative humidity is high. As a rule of thumb, the RPE is 1.5 to 2.5 Borg scale categories (1 to 2 on the OMNI scale) above the normal training zone when the air temperature is hotter than 88° F (Pandolf 2001). Clients who have an elevated RPE under these conditions should reduce their exercise intensity in order to avoid heat strain. They will make this adjustment almost automatically if they are reminded to exercise at a slower pace on hot days and when the relative humidity is above 70% (Pandolf 2001). Reassure clients that although their pace will be slower, they will still be in their prescribed RPE training zone.

Research has shown that RPE is related in part to the client's sensation of temperature, that is, how hot they feel (Pandolf 2001). A special OMNI scale was developed to facilitate the measurement of temperature sensation (see appendix A). The temperature sensation categories on the scale increase from *neutral* (1) to *very hot* (5). The scale's pictures of a progressively larger sun help clients to rate how hot they feel. This modified scale can be used in conjunction with the standard OMNI scale format to simultaneously measure temperature sensation and RPE. The upper-limit alarm ratings on the OMNI scale for exercise under hot or humid conditions are an RPE of 8 and a temperature sensation of 4. If the client gives both an upper-limit RPE and an upper-limit temperature-sensation rating, reduce the training pace until both ratings fall below their upper-limit levels.

A variant of this concept measures RPE and thermal sensation to compute the Perceptual Heat-Strain Index, or PeSI (Tikuuisis, Mclellan and Selkirk 2002).

$$\text{PeSI} = \left[5 \cdot \frac{(\text{TS} - 7)}{6}\right] + \left[5 \cdot \frac{\text{PE}}{10}\right]$$

where *TS* is a thermal sensation of 7 to 13 on the Gagge Scale (ratings ranging from cold to hot) and *PE* is perceived exertion on the Borg CR-10 Scale.

This perceptually based index is accurate for untrained clients, but trained athletes tend to underestimate their perceived heat strain. Therefore, the PeSI should be used in conjunction with physiological measures (such as the core temperature) to assess heat strain in competitive athletes.

In general, as clients acclimatize to a high air temperature, RPE decreases for a given exercise intensity. This indicates that the HR and body temperature have decreased and the plasma volume and sweat rate have increased in response to repeated exposure to hot environmental conditions (Pandolf 2001). Acclimatization usually requires 3 to 6 consecutive days of exposure to hot air conditions. As the RPE decreases with acclimatization, exercise intensity must be increased to maintain the appropriate target training zone.

Cold Temperature and RPE

Aerobic exercise performed in cold air—such as cross-country skiing and winter jogging—produces an RPE of approximately 1 Borg scale (0.5 OMNI scale) category lower for a prescribed training pace. This drop in RPE normally does not occur until the outdoor temperature is below 52° F.

High Altitude and RPE

When exercise is performed at high elevations (usually above 4000 ft), the RPE can be as much as 3 Borg scale (2 OMNI scale) categories above normal (Maresh et al. 1993). To maintain the optimal RPE training zone, clients will likely need to reduce their training pace when they condition at high elevations.

OCCUPATIONAL APPLICATIONS OF RATINGS OF PERCEIVED EXERTION

The term **ergonomics** refers to the use of human factors engineering to assess perceptual, physiological, and physical aspects of a work task

(Borg 1998). These measures of the *worker's cost* take into account the time required to do the task, the frequency of the task's performance, and the difficulty of the task. The objective of ergonomic assessment is to achieve a balance between the worker's cost and the quality and quantity of the worker's output by modifying the work task so that it fits the worker's perceptual and physiological capacity. This *fit* is especially important for workers who must perform physically demanding tasks during some or all of their shift. Such jobs as fire fighting, refuse disposal, lumber work, commercial fishing, steel manufacturing, and farming require high levels of energy expenditure at equally high levels of perceived exertion. Despite industrial automation, occupationally related injuries and illnesses, especially those associated with repetitive motion, remain prevalent throughout the workforce.

Two factors establish the worker's cost—the energy required to perform the task and the level of exertion perceived when performing that task (Robertson and Noble 1997). Assessing the energy cost of an occupational task is expensive, technically difficult, and disruptive. However, energy requirement and RPE are closely linked (see chapter 1). Therefore, it is practical to estimate the worker's cost by assessing RPE during and sometimes immediately after completion of the work task.

A number of different RPE scales can be used in occupational settings. A scale that was specifically developed to assess RPE in the workplace is the Occupational Effort Index (Hogan and Fleischman 1979). This index has numerical categories ranging from 1 to 7, each of which is linked to a verbal cue. The verbal cues are the same as those that appear in the original 6-20 category Borg RPE scale. The Occupational Effort Index has been validated using energy cost measurements for 24 different manual tasks. When averaged over all tasks, the validity correlation between RPE and energy cost was $r = 0.88$.

Several other category scales of perceived exertion can also be used in the workplace. Among these are the Borg 6-20 category and Borg CR-10 scales. Although the OMNI scale has not been employed as extensively in this setting, its easy-to-understand format makes it a useful tool for assessing the strain that workers experience while performing a physically demanding task.

Measuring RPE in the occupational setting provides a subjective reference that reveals how the job can be modified to fit the worker. In general, the job tasks that have the lowest RPEs and energy costs are considered ergonomically optimal for the worker. For example, a worker who carries a load equal to 7.5% of her body weight has neither increased energy expenditure nor RPE compared to walking without a payload (Robertson et al. 1982). However, when a worker transports a payload heavier than 7.5% of her body weight, both energy expenditure and RPE are significantly elevated. Therefore, it is recommended that the weight of a manually transported payload not exceed 7.5% of the worker's body weight. These ergonomic guidelines make it clear that payload weight has to be adjusted to account for the general differences in body weight between female and male workers. RPE measurement has also been applied in determining the appropriate style for lifting a payload. In general, for a given payload weight, RPE is lower when a squat style as opposed to a deadlift style is employed (Gamberale et al. 1987). Adjusting the characteristics of a work task using RPE as a guide is easy and helps to prevent job-related injuries and illnesses.

SUMMARY

This chapter explains how to use RPE zones to guide training programs for special groups, including women, children, and older individuals. These uniquely designed exercise programs use RPE conditioning zones as a simple and effective method for guiding exercise intensity. Training programs and target RPE zones are also presented for clients who have such clinical conditions as osteoporosis and pregnancy. Examples of how sliding RPE zones can be used to guide children's physical activity are given. Finally, RPE can be used to monitor environmentally induced thermal strain and to establish ergonomically acceptable work conditions so that they fit the worker's perceptual and physiological capacities.

C·H·A·P·T·E·R

8

Rating of Perceived Exertion Training Zones for Competition

CASE STUDY

Client Characteristics

The athlete is a 20-year-old male who has been competing in high school and college track events for 4 years. He specializes in middle- and long-distance events. He enjoys training and competing and would like his abilities to approach championship level during the upcoming outdoor track season. His coach has promised to give him more individualized training to help him meet this objective.

Exercise Need

The athlete's coach uses quantity and quality interval formats (discussed later in this chapter) extensively during the in-season training period. Because the track team is quite large, the athlete trains with a group of middle- and long-distance runners. The coach regulates both the intensity of the exercise intervals and the length of the recovery period between intervals, using standard times for all the runners in the group. However, because performance ability varies markedly between the more- and less-experienced runners in the group, it is not possible to individualize the stimulus overload according to training status and fitness level using this procedure.

Action Plan

The coach decides to regulate both the exercise and recovery intervals by using target RPEs. The intensity of the quantity and quality intervals slides between OMNI RPE training zones 8 and 9. The length of the recovery interval is set at the time required for each runner's RPE to return to 4 or 5 on the OMNI scale. Therefore, the time to perform the interval and the time allotted for recovery are individualized for each runner in the training group.

onditioning for competitive athletics involves both core and sport-specific training. The goal of **core training** is to improve the overall levels of aerobic and anaerobic power and muscular strength and endurance. Core training provides a general level of fitness that serves as a foundation on which sport-specific skills are built. **Sport-specific training** programs build on the core foundation to enhance skill and game performance. An example of sport-specific training is the use of a resistance exercise to increase the arm strength of a baseball pitcher where the lifting action simulates throwing a baseball. RPE zones can be used to guide both core and sport-specific training. To address the broadest possible applications of RPE training zones for individual and team sports, this chapter primarily examines core training programs. However, several sport-specific training examples are also considered.

OVERLOAD TRAINING PRINCIPLE

Both core and sport-specific conditioning programs employ the overload training principle. **Overload training** gradually, progressively, and systematically applies the training stimulus to enhance both physiological and psychological determinants of sport performance. The training stimulus actually involves three different exercise loads—a normal load, peak load, and overload (Hoffman 2002). A **normal training load** allows the athlete to achieve and maintain a *performance steady state* for a specified period during the workout. A **peak load** is the greatest normal load that can be managed. Any training stimulus that is greater than the peak load is an **overload.** These three exercise loads are applied within a periodized training progression divided into macrocycles, mesocycles, and microcycles (Kreider, Fry, and O'Toole 1998). **Macrocycles** generally are 3 to 4 months in duration, roughly corresponding to the off-, pre-, and in-season periods of training and competition. A single macrocycle is divided into a sequence of **mesocycles,** each one approximately 4 weeks in duration. Within a given mesocycle are a series of 1- to 7-day-long **microcycles.** It is within these microcycles that the training stimulus is alternated gradually, progressively, and systematically between peak loads and overloads, with brief periods for rest and regeneration interspersed as needed.

QUANTITY AND QUALITY OVERLOAD TRAINING

An overload training program can use either a quantity or quality format. A **quantity training format** is appropriate for distance (endurance) events that require high levels of aerobic energy and hence cardiorespiratory fitness. The goal of quantity-training programs is to increase the pumping capacity of the heart to improve the central circulatory transportation of oxygen to exercising muscles. A **quality training format** prepares athletes for speed and power events and has two physiological targets. First, it enhances the capacity of the trained peripheral skeletal muscles to extract oxygen from the blood and in turn to oxidize fat and carbohydrates for energy. Second, it increases the capacity of trained muscles cells to anaerobically metabolize glycogen for energy.

Within a given macrocycle, the overload stimulus should always follow a quantity-to-quality training progression. The first step in selecting a quality, quantity, or combined overload training format is to partition the aerobic and anaerobic energy requirements of the sport. For example, the total energy required to perform a marathon is composed of 90% aerobic and 10% anaerobic energy. An 880 yd run, on the other hand, requires 5% aerobic and 95% anaerobic energy. Table 8.1 lists the energy demands of competitive events that range from almost all aerobic to almost all anaerobic energy. The next step in designing the exercise program is to select the overload training format that has the appropriate mixture of quantity and quality training to meet the specific energy demands of competition. Four core overload training formats are specific to the quantity and quality energy demands of most individual and team sports: continuous, fartlek, interval, and repetition training (table 8.2).

A **continuous format** employs long, slow over-distance training at 70 to 80% of maximum effort for workouts of 30 to 60 min in duration. *Fartlek* is a Swedish term meaning *speed play*. In **fartlek training,** the intensity is alternated between slower (75 to 80% of maximum) and faster (80 to 85%) bouts, with the total workout lasting 30 to 60 min. The fartlek format is

Table 8.1 Energy Partition According to Performance

Run distance	Speed (anaerobic)	Endurance (aerobic)
Marathon	10%	90%
6 miles (10,000 m)	20%	80%
3 miles (5,000 m)	30%	70%
2 miles (3,000 m)	60%	40%
1 mile (1,500 m)	75%	25%
880 yd (800 m)	95%	5%

Adapted from J. Hoffman, 2002 *Physiological aspects of sport training and performance* (Champaign, IL: Human Kinetics).

Table 8.2 Core Training Formats

Format	Intensity (% $\dot{V}O_2$max)
Continuous	70–80
Fartlek	75–80 and 80–85

Interval
   Quantity
   Quality
	
	85–90
	90–95

Repetition	95–100

basically quantitative continuous exercise interspersed with brief qualitative bouts called *speed play.* The **interval training** format has four components: (a) the speed of the interval, (b) the number of intervals, (c) the duration of the intervals, and (d) the length of the recovery time between exercise intervals. Any one or a combination of these components can be altered within a given interval workout to provide a training overload. The format is considered either quantitative or qualitative depending on the speed of the exercise interval. A quantity interval is performed at 85 to 90% intensity, whereas a quality interval is performed at 90 to 95%. The most important component of the interval format is the recovery period between exercise intervals. The period should only be long enough to allow 50% recovery between intervals. This provides an overload stimulus not only during the exercise interval, but also during the recovery period. To accomplish this, the recovery period should never be longer than 50% of the time required for the exercise interval. The fourth core format is called **repetition training.** In this format, the exercise repetition is performed at or at near race pace (approximately 95 to 100% of maximum). Because the intensity is near the maximum, the period after each exercise repetition should be sufficient to allow complete recovery prior to starting the next repetition. However, unnecessarily long recovery periods limit the total number of repetitions that can be performed in a given workout. To avoid this problem, limit the distance of each exercise repetition to approximately 50% of the actual race distance. The repetition format provides a quality overload training stimulus.

Figure 8.1 shows in a diagram how the four core formats are used in a progressive, year-round training program consisting of three macrocycles—off-, pre-, and in-season. Because the four core formats are undertaken sequentially, the year-round macrocycles provide for a quantity-to-quality progression of the overload training stimulus. The progression begins with the off-season macrocycle, when continuous training is the primary format. As the off-season draws to a close, fartlek

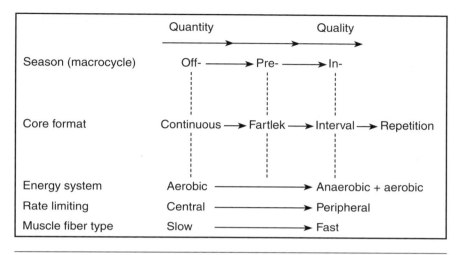

Figure 8.1   Quantity-to-quality concept in year-round cardiovascular training.

training is begun. This adds a qualitative dimension to the quantitative base established with continuous training in the off-season. The fartlek format is a transition period that shifts from the predominantly quantity off-season training to a mix of quantity and quality preseason training. Once the preseason macrocycle begins, quantity intervals are added to the mix of previous formats. During in-season competition, quality intervals and repetition formats are also added to the training mix. The quality-interval format is used as the primary method to produce the overload training stimulus. Periodically throughout the competitive season, repetition training is introduced. A repetition workout is usually prescribed just before or during the pre-event tapering period to let the athlete practice in simulated competitive conditions as a lead-in to the actual event. As depicted in figure 8.1, the four core formats provide an overload stimulus that always progresses from quantity to quality, central to peripheral, and aerobic to anaerobic dimensions in a year-round training program. The key is to establish and reinforce the quantitative training base while progressively adding qualitative refinement to the conditioning program.

CORE RATING OF PERCEIVED EXERTION TRAINING ZONES

We have examined how the core training formats provide a quantity-to-quality overload training progression. The next step is to determine how the intensity of each core format is guided by an RPE training zone. The intensity of the RPE zone increases progressively from the continuous to fartlek to interval to repetition training formats. The RPE zone for each core training format is listed in table 8.3. An RPE zone can also be used to regulate the recovery period between exercise intervals. As noted earlier, the length of the recovery period should not be more than 50% of the actual exercise time for the interval. The RPE zones typically associated with a 50% recovery point are 4 to 5 on the OMNI scale and 11 to 13 on the Borg 6–20 Sscale (Robertson et al. 1992). Therefore, the recovery RPE can be used in conjunction with the amount of time elapsed to determine when the next interval should be started.

Table 8.3   RPE Zones for Core Training Formats

	Core formats				
			Interval		
RPE zone	Continuous	Fartlek	Quality	Quantity	Repetition
OMNI (0–10)	6	6 and 7	8	9	10
Borg 6–20	14	14 and 16	17	18	20

RATING OF PERCEIVED EXERTION TRAINING ZONES FOR RESISTANCE EXERCISE

Progression through a year-round resistance exercise training program for competitive athletes can be guided by the cycles described above (Wathen and Roll 1994). Each cycle emphasizes a different type of muscular development. The off-season macrocycle initially focuses on muscular endurance (high volume, low intensity) and gradually shifts to muscular strength (low volume, high intensity). The preseason macrocycle develops muscular endurance (high volume, low intensity) with sport-specific exercises. Finally, the in-season macrocycle consists of low-volume, high-intensity, position-specific exercises. The sliding RPE zone system can be used with the OMNI-RES to guide the training intensity during each of these three macrocycles. An RPE training zone of 3 promotes muscular endurance, a zone of 6 encourages muscular hypertrophy, and a zone of 9 develops muscular strength. How to use these resistance exercise training zones is described in chapter 6.

RATING OF PERCEIVED EXERTION INDICES OF OVERTRAINING SYNDROME

Overtraining is caused by a combination of psychological and physiological maladaptive responses to an inappropriate training stimulus. Because RPE is a robust psychophysiological marker of the strain imposed on an athlete during practice and competition, it acts as a barometer, warning of impending overtraining (Foster et al. 1995; Svedenhag and Seger 1992).

First, it is necessary to distinguish between overreaching and overtraining (Kreider, Fry, and O'Toole 1998). **Overreaching** is a short-term performance decrement that occurs despite a continued increase in the training overload. The downturn in performance can occur with or without clinical, physiological, or psychological signs or symptoms of training maladaptation. As a general rule, overreaching can be reversed by inserting short rest and regeneration periods into a normal training microcycle. When this is done, the overreaching state may actually provide the adaptational foundation for improved performance. In contrast, **overtraining** is revealed by a profound, long-term performance decrement typically accompanied by signs and symptoms of excessive exercise strain. Overtraining can be reversed only through mandatory and prolonged periods of rest and regeneration, often followed by low-volume cross-training (Kreider, Fry, and O'Toole 1998). It is important to note that intentional overreaching, when inappropriately applied, can lead to overtraining.

Because the reversal of overtraining is a long-term process that often requires the athlete to be removed from both training and competition, it is of paramount importance to prevent it by quickly responding to downturns in performance. The most effective method for preventing overtraining is to intersperse rest microcycles throughout each mesocycle. Alternatively, some coaches prefer using low-volume (reduced in intensity and duration) microcycles rather than rest microcycles. Normally, these rest microcycles are scheduled in advance. If the early warning signs of overtraining are noticed, however, a rest or low-volume microcycle should be started immediately.

RPE responses play a key role in tracking the onset, severity, and eventual reversal of both overreaching and overtraining. An effective overload training program for high-level competition should push athletes' physiological and psychological reserves up to but not beyond the border between positive performance adaptation and overtraining maladaptation. Because the use of RPE overtraining indices is particularly helpful in identifying the first signs of overtraining, a number of practical RPE indices have been developed to signal overtraining and to track its reversal after appropriate rest and regeneration microcycles have been initiated. Note that these indices are based primarily on laboratory studies; field-based validation is pending. A brief description of three of them follows.

RPE Monotony Index (Foster et al. 1995)

- Measure the overall (i.e., global) RPE using the Borg C-R 10 scale approximately 30 minutes after a given training session. This global rating is called the session RPE.
- Record the duration (min) of the entire training session (i.e., warm-up, cool-down, and recovery).
- Multiply the session RPE by the session duration. This is called the session load.
- Calculate the session load for five to seven consecutive training sessions.
- Next, calculate the mean and standard deviations of these five to seven session loads.
- The Monotony Index is calculated by dividing the mean by the standard deviation.
- If the Monotony Index consistently increases over several measurement periods, regular monitoring for signs and symptoms of overtraining should be initiated.

Lactic Acid:RPE Ratio (Snyder et al. 1993)

- Periodically measure the athlete's RPE (OMNI or Borg scale) and blood lactic acid ($mm \cdot L^{-1}$) responses to a set submaximal training pace.
- Divide the amount of lactic acid by the RPE and multiply that value by 100.
- If the ratio consistently decreases over several measurement periods, regular monitoring for signs and symptoms of overtraining should be initiated.

Perceptual-Mood Index of Overtraining (modified from Morgan et al. 1988)

Perceptual-Mood Index (PMI) = [(RPE-O + Fatigue)/Vigor] $\times$ 100

- where RPE-O is the overall body rating of perceived exertion on the Borg 6–20 Scale. Fatigue and vigor values are determined with the Profile of Mood States.
- Measure RPE-O and the fatigue and vigor mood states at scheduled times during a mesocycle.
- Calculate the PMI as shown previously.

- If the PMI has increased, it indicates that overtraining is approaching. For example:

$$\text{Normal PMI} = [(14 + 7)/21] \times 100 = 100$$
$$\text{Overtrained PMI} = [(18 + 18)/13] \times 100 = 276$$

An RPE overtraining index should be administered at the beginning of the off-season macrocycle to establish a baseline index value for each athlete. The index can then be administered periodically during the various mesocycles to monitor for evidence of overtraining by comparing the current RPE index with the off-season baseline value. Often, the RPE index will reveal imminent overtraining before practice or competitive performance declines do. The RPE indices are especially useful during the busy in-season macrocycle, when the training schedule of athletes preparing for top-level competition is often very full. Collecting the RPE data and calculating the indices requires less than 5 min and does not disrupt training.

Athletes can be tested for overtraining at very high exercise intensities. During short-term treadmill running at a very high intensity (such as 125% $\dot{V}O_2$max), the initial RPE is less than 20 on the Borg 6–20 Scale (Doherty et al. 2001). The RPE rises as exercise continues and when it reaches 20, the test is terminated. Measuring RPE during very high intensity exercise is helpful in tracking overtraining symptoms. Often, the earliest signs of overtraining occur during quality workouts at high intensities, when both physiological and psychological stress are greatest.

Periodic tracking with RPE overtraining indices is a valuable tool for preventing overtraining. Prevention is vitally important because, as a general rule, it takes much less time to reverse a trend toward overtraining than to "rehabilitate" overtrained athletes. When using RPE to track overtraining, the athlete's own *Percepto-Stat* can be of assistance. Morgan and Pollock contend that endurance athletes are of two different types—associative and dissociative—in their cognitive approach to training, which the researchers call their *Percepto-Stat* (1977). *Associators* focus on both their internal perceptual cues and their external performance environment at all times during training and competition. In doing so, they are tuned in to their Percepto-Stat, which allows them to make constant adjustments to their race strategy based on internal and external exertional signals. Athletes who are tuned in to their Percepto-Stat can also adjust their exercise levels to meet a training overload but avoid overtraining and overuse injury. *Dissociators,* in contrast, do not pay attention to internal or external cues and instead listen to music, fantasize, or solve problems during exercise. Their Percepto-Stat is turned down, often predisposing them to overtraining and injury.

TRACKING TRAINING PROGRESS

RPE can be used to track progress during a year-round training program. Changes in RPE over the course of the season or a specific training cycle indicate that (a) there is improvement or deterioration in fitness level, (b) the training dosage must be adjusted to accommodate fitness changes, and (c) the athlete may be overtraining (Noble and Robertson 1996). The two types of RPE tracking are constant level tracking and constant RPE tracking.

Constant level tracking is based on our knowledge that as the athlete's fitness level changes during the course of the training cycle, the RPE response at a standardized exercise level also changes. In constant level tracking, the athlete performs at a set exercise pace periodically throughout the training cycle. This is usually done during the first 5 to 10 min of a workout. The constant level exercise can be cycling at a specific submaximal PO on a stationary ergometer, running or swimming at a particular speed for a given distance, or lifting the same amount of weight. As athletes' fitness increases over the training cycle, their RPE at the fixed exercise level decreases. If fitness has decreased or overtraining is about to occur, the RPE increases. In the former case, the training dosage (either intensity or duration) should be increased, but when overtraining is imminent, the dosage should be examined to determine if it is inappropriately high. Adjustments can then be made in the dosage and a strategy for employing the optimal mixture of training and rest microcycles can be introduced.

Pierce, Rozenek, and Stone reported on their use of constant level tracking in a resistance training program (1993). They obtained tracking responses from healthy young men who were participating in a resistance training program. The participants lifted a weight equivalent to approximately 63% 1RM on the first day of the training program and then measured their RPE-AM on the Borg 6–20 Scale. They again lifted the same absolute weight at the end of the training program. The average RPE-AM decreased from 18 at the initial test to 12 at the final test. In conjunction with this decrease in RPE, muscular strength in the trained muscles increased by 23%. Therefore, the decrease in the RPE during the constant level tracking test signaled a positive training adaptation. If the participants had chosen to continue training, the workout dosage would have had to be increased until the original target RPE was again produced. This tracking system is a convenient, technically simple method for monitoring training progress that has applications in both aerobic and resistance exercise programs.

Constant RPE tracking is based on our knowledge that as fitness changes, the exercise level (intensity) has to change in the same direction to produce the same RPE. For example, at the beginning of a training season, athletes perform a standard 5-min exercise bout at a fixed, submaximal PO on a cycle ergometer and then measure their RPE. This rating is called the *target tracking RPE*. Periodically throughout the training season, the athletes perform the same submaximal test, but are instructed to change the PO until they produce the target tracking RPE. If the new PO at the tracking RPE is greater, the fitness level has increased. Conversely, if the new PO is less, the fitness level has decreased. A lower PO may be an early indication of overtraining.

When using either of these tracking systems, it is important that the muscle groups that are activated in the tests are the same as those worked in the training program. This significantly enhances the accuracy of the RPE tracking system and ensures that subsequent adjustments in the exercise dosage are specific to training adaptations in the activated muscle groups.

SUMMARY

This chapter describes the use of RPE training zones by competitive athletes. Overload training employs gradual, progressive, and systematic application of a training stimulus. The overload stimulus is applied using a sequence of formats that progress from quantity to quality training—continuous, fartlek, interval, and repetition. The intensity of each format is regulated according to sliding RPE training zones that are effective for both aerobic and strength conditioning. Using RPE indices to track training progress and signal impending overtraining is also discussed.

C·H·A·P·T·E·R

9

Exercise Testing and Training for the Rehabilitation Patient Using Ratings of Perceived Exertion

CASE STUDY

Client Characteristics

The client is a 50-year-old male who has a 5-year history of mild chronic obstructive pulmonary disease (COPD) and is regularly followed by a pulmonary physician. He has no cardiac, metabolic, or orthopedic exercise limitations. The client performs normal ADLs without restriction, is employed full-time, and enjoys outdoor recreational activities.

Exercise Need

Recently, the client has experienced exertional dyspnea when gardening and biking. To reduce these symptoms, his physician has prescribed an aerobic exercise program. He is currently performing the exercise program at a health-fitness club 3 to 5 days per week under the supervision of a clinical exercise practitioner. The practitioner needs a quick, simple way to track the client's training progress that will not interfere with the client's exercise session.

Action Plan

The clinical practitioner measures the client's Dyspnea Severity Index for COPD (Mahler and Horowitz 1994) during the first 5 min of the training session on a biweekly basis. The dyspnea index (Borg CR-10 dyspnea rating ÷ watts) is measured during a submaximal cycle ergometer protocol that uses the same PO at each testing session. For a given submaximal PO, the index decreases as the client's dyspnea improves with exercise training. The index provides both client and practitioner with information regarding therapeutic progress and the adequacy of the exercise prescription.

RPEs can be used to assess exercise tolerance and prescribe exercise rehabilitation programs for both cardiac and pulmonary patients. These exercise applications must always be positioned within the functional and symptom-limited constraints of the patient's clinical status.

EXERCISE THERAPY
FOR THE CARDIAC PATIENT

Exercise rehabilitation for the cardiovascular patient is normally divided into inpatient and outpatient programs (ACSM 2000). **Inpatient** rehabilitation assesses the individual's tolerance for the aerobic and strength requirements of daily living and provides lifestyle education and counseling. **Outpatient** cardiovascular rehabilitation promotes total physical conditioning with multiple activities, including aerobic, range-of-motion, and resistance exercise. Activities should be chosen to maximize the transfer of training benefits to real-life settings. RPE can be used to regulate the intensity of exercise therapy for both inpatient and outpatient cardiac rehabilitation programs (Noble and Robertson 1996).

Perceived exertion can be used in conjunction with HR to regulate exercise intensity during inpatient therapy. The upper-limit target HR for inpatient ambulatory activities should be 20 beats $\cdot$ min^{-1} above the patient's HR at standing rest. The corresponding RPEs for this intensity are 8 to 10 on the Borg 6–20 Scale and 2 to 3 on the OMNI scale. This RPE zone is appropriate for most inpatient ambulatory activities, but neither an HR nor an RPE that produces ischemic symptoms should be chosen as a target. The upper-limit HR and RPE should be modified according to the patient's clinical and perceptual responses during the first several exercise sessions.

The physiological and clinical assumptions that underlie the prescription of exercise for outpatient management of the coronary patient are— with one notable exception—the same as those for the clinically normal individual. The exception is that exercise intensity must be set within the functional or clinical constraints imposed by the patient's ischemic threshold. From a physiological standpoint, the **ischemic threshold** is reached when myocardial oxygen demand exceeds the supply. Clinical signs and symptoms of exercise-induced ischemia include electrocardiogram abnormalities, angina, inappropriate blood pressure responses, dizziness, unsteady gait, and sudden pallor and profuse sweating. These responses to exertion are distinct from the normal signs of exertional fatigue that occur during the exercise session.

Like that for a healthy individual, the objective of an exercise prescription for the coronary patient is to achieve a total body oxygen consumption level that falls within a predetermined stimulus zone to improve cardiorespiratory fitness. However, the cardiac output needed to meet a prescribed total body oxygen uptake must not require the left ventricle to work so hard that myocardial oxygen supply is compromised. When the oxygen supply is inadequate for the myocardial workload, the ischemic threshold is exceeded, producing symptoms of coronary insufficiency and exertional intolerance. The physiological and clinical characteristics of the ischemic threshold are identified with a GXT. Once identified, the ischemic threshold serves as a prescriptive reference on which the intensity of exercise is based. This is normally accomplished by selecting a therapeutic training zone that is equal to or approximately 1 metabolic equivalent (MET) below the ischemic threshold (ACSM 2000). It should be noted that 1 MET equals a $\dot{V}O_2$ of 3.5 ml/kg/min. The RPE that corresponds to the therapeutic training zone is then identified from a plot that expresses perceived exertion as a function of METs (see chapter 5). These RPE and MET responses are determined during a preparticipation GXT. As a general rule, most outpatient cardiac rehabilitation falls within an RPE zone of 11 to 13 on the Borg scale and 4 to 5 on the OMNI scale. The target RPE can then be used to keep the exercise intensity within a clinically safe training zone that does not exceed the ischemic threshold. Tables 9.1 and 9.2 list the therapeutic RPE zones that are recommended for cardiac rehabilitation programs that include aerobic and resistance exercise.

Table 9.1   Aerobic RPE Zones for Cardiac Rehabilitation

Program component	Functional training intensity*	Therapeutic RPE zone	
		OMNI 0–10	Borg 6–20
Warm-up and cool-down	20–40%	1–2	7–8
Low stimulus	40%	2–3	9–10
Moderate stimulus	60%	4–5	11–12
High stimulus	80%	5–6	13–14

* Percent of HR range.

Adapted, by permission, from M.H. Whaley et al., 1997, "Validity of rating of perceived exertion during graded exercise testing in apparently healthy adults and cardiac patients," *Journal of Cardiopulmonary Rehabilitation* 17: 261-267.

Table 9.2   Strength Training for the Cardiac Rehabilitation Patient Using OMNI RPE Zones

Week	Exercise	Muscles	Sets	OMNI RPE zone*
1	Machine decline press	Pectoralis major, deltoids, triceps	1	3
	Machine seated row	Latissimus dorsi, biceps, shoulder retractors	1	3
2	Machine leg extension	Quadriceps	1	3
	Machine leg curl	Hamstrings	1	3
3	Machine back extension	Erector spinae	1	3
	Machine abdominal curl	Rectus abdominis	1	3
4–7	Continue to perform 1 set of all 6 exercises at OMNI RPE zone 3.			

During and after week 8, add 1 set each of machine hip adduction and abduction (hip adductors and abductors), dumbbell lateral raise (deltoids), dumbbell arm curl (biceps), dumbbell arm extension (triceps), and dumbbell shrug (upper trapezius). Guide target training zones with the OMNI-RES as described in chapter 6.

From W.L. Westcott and S. O'Grady, 1998, "Strength training and cardiac postrehab," *IDEA Personal Trainer*. Reproduced with permission of IDEA Health & Fitness Association, (800) 999-IDEA, www.IDEAfit.com.

* Added by R.J. Robertson as a supplement to this table.

EFFECT OF CARDIAC MEDICATIONS

Medications such as β-blockers and cardiac stimulants are routinely used by patients undergoing cardiovascular exercise rehabilitation. Many of these cardioactive medications alter the normal HR response to exercise. For example, β-blockers such as propranolol attenuate exercise HR, whereas cardiac stimulants such as atropine accelerate it. Regulating exercise intensity according to a target HR is confounded when the dosage of these medications is periodically adjusted to achieve optimal therapeutic benefit. However, exertional perceptions are generally stable in the presence of cardioactive medications (Eston and Connolly 1996). Despite alterations in exercise HR that can occur with medication dosage changes, patients can maintain the prescribed intensity by exercising to a target RPE zone.

β-blockers are also used to treat migraine headache, anxiety, and tremor (Eston and Connolly 1996). Therefore, many physically active young and middle-aged individuals also take these agents. At a given sub-maximal exercise intensity, both nonselective and selective β-blockade therapy can elevate RPE, especially when the rating is differentiated to the exercising muscles. This is because maximal levels of oxygen consumption and cardiac output are usually reduced by β-blockade, which makes a given submaximal intensity relatively greater and more fatiguing. However, when a training zone is based on a % $\dot{V}O_2$max, the target RPE is not substantially altered by β-blockade. Most patients undergoing β-blocking therapy can effectively and safely self-regulate their therapeutic exercise intensity using a target RPE zone.

RATING OF PERCEIVED EXERTION THERAPEUTIC ZONES FOR GROUP EXERCISE

The prescription of a therapeutic exercise intensity using a target RPE zone is very useful when cardiac rehabilitation is performed in group training sessions. Outpatient rehabilitative activities correspond to an RPE of 11 to 13 on the Borg 6–20 Scale and 4 to 5 on the OMNI scale. When the HRs corresponding to these target RPE zones are expressed as % HR_{max}, the training stimulus is fairly similar in patients who have mild and more advanced disease (Noble and Robertson 1996). This facilitates supervision of cardiac exercise rehabilitation programs in which groups of patients with varying disease severity participate at the same time.

RATING OF PERCEIVED EXERTION AND CARDIAC TRANSPLANT PATIENTS

The peak HR response during dynamic exercise is markedly blunted in the denervated transplanted heart. Therefore, HR is not a sensitive measure for tracking the therapeutic progress of transplant patients participating in cardiac exercise rehabilitation. In contrast, RPE is an effective and easy-to-use tool for this purpose. For example, Shephard and colleagues (1996) demonstrated that after 16 months of exercise rehabilitation, $\dot{V}O_2$max increased by 19% in a group of cardiac transplant patients. When these patients were tested at a given submaximal PO and $\dot{V}O_2$, RPE was significantly lower at the post-training than the pre-training assessment. The decrease in RPE over the course of the rehabilitation program indicated that improvement had occurred in symptom-limited functional aerobic power.

In general, a target training RPE zone of 13 on the Borg 6–20 Scale or 5 on the OMNI scale is appropriate for most transplant patients when used in conjunction with a standard time per distance exercise prescription.

The therapeutic training progress of cardiac transplant patients can easily and accurately be tracked by using RPE responses to a submaximal cycle ergometer protocol.

RATING SCALES FOR PULMONARY DISEASE AND PAIN

Category rating scales for the clinical assessment of dyspnea and pain can be used as companions to the OMNI and Borg perceived exertion scales. Most patients can distinguish between sensations caused by exercise and those associated with disease processes. Therefore, it is common practice in clinical settings to use multiple scales to assess sensations that reflect both disease severity and level of physical exertion. A therapeutic exercise prescription for coronary and pulmonary patients can then be developed in which the target training RPE zone reflects exertional limits imposed by both disease severity and functional fitness levels. Such a prescription is both medically safe and physiologically effective (Noble and Robertson 1996).

Exercise diagnosis of lung disease and subsequent classification of disease severity can be facilitated by category rating scales that use a basic RPE format. Three different indices for evaluating the clinical status of pulmonary patients have been developed using the Borg CR-10 Scale. They are listed in table 9.3 and explained in the following sections.

Table 9.3   RPE Indices of Pulmonary Disease and Dyspnea Based on the Borg CR-10 Scale

Index*	Calculation
COPD severity	(dyspnea rating** ÷ w) × 100
Respiratory rate (RR) index	RPE plotted at RR**
Rating of perceived breathing difficulty	Select 0–10**

* See text for explanation. ** Borg's CR-10 scale.

DYSPNEA INDEX FOR CHRONIC OBSTRUCTIVE PULMONARY DISEASE

Mahler and Horowitz (1994) developed a severity index for COPD. During an exercise test, COPD patients rate their feelings of dyspnea (shortness of breath) on the Borg CR-10 Scale. A *dyspnea rating* of zero indicates no breathing difficulty and a 10 indicates maximum breathing difficulty. The dyspnea rating is measured for each PO stage during a load-incremented cycle ergometer exercise test. The dyspnea index is calculated as the ratio of the dyspnea rating to PO in watts—(CR-10 rating ÷ w) × 100. At any stage during the exercise test, the higher the index value, the more severe is the patient's obstructive lung disease. At a moderate cycle ergometer PO such as 75 W, patients with severe COPD (FEV_1 < 50% predicted) have a dyspnea index of approximately 8, whereas patients with mild to moderate COPD (FEV_1 50 to 70% predicted) have an index of approximately 6. The dyspnea index can also be used to track the therapeutic training progress of COPD patients who are participating in pulmonary exercise rehabilitation. For a given submaximal PO, the index decreases as the patient's dyspnea improves with exercise training.

RPE–Respiratory Rate Index

An easy-to-apply procedure for tracking the therapeutic progress of pulmonary patients uses measures of respiratory rate (RR) and RPE on the CR-10 scale (Linderholm 1986). Both RR and RPE are assessed during each minute of a load-incremented cycle ergometer exercise test. The RR is then plotted against the corresponding RPE. The test is administered periodically over the course of a therapeutic exercise program. At a given RR, the RPE will be lower as the patient's clinical status improves.

Rating of Perceived Breathing Difficulty

A modified Borg CR-10 scale can be easily and effectively used to assess exertional dyspnea (Covey et al. 2001). The pulmonary patient is asked to rate how difficult it feels to breathe while exercising. This response, termed a *rating of perceived breathing difficulty* (RPBD), is a valid clinical tool that can be used to assess exertional tolerance and to track therapeutic progress in pulmonary patients who are participating in a rehabilitation program. Clinical progress is indicated by a reduction of approximately 2 RPBD units from one test period to the next. The following instructions should be read to the patient when using this index (Franco et al. 1998).

> This is a scale for rating breathlessness. The number zero represents no breathlessness. The number 10 represents the strongest or greatest breathlessness you have ever experienced. Each minute during the exercise test, you will be asked to point with your finger to the number that represents your perceived level of breathlessness at that time. Use the written description to the right of the number to help guide your selection. I will say the number out loud to confirm your choice.

> During the exercise, you may have an even stronger or greater intensity of breathlessness than you have ever experienced. You should then point to the word *maximal* if the severity is greater than 10. You can tell us this rating after the mouthpiece has been removed.

As the patient improves with exercise therapy, the RPBD decreases, indicating that ventilatory muscle strength and endurance have improved. However, dyspnea relief is also linked to a reduction in fear and anxiety, increased exercise tolerance, and systemic desensitization to respiratory discomfort. Clinical practitioners should keep in mind that these beneficial subjective responses to exercise rehabilitation can occur even if measurable physiological adaptation is not present.

RATING OF PERCEIVED EXERTION ZONES FOR PULMONARY REHABILITATION

Pulmonary patients can safely and effectively regulate the intensity of an exercise rehabilitation program by using a target RPE zone that accommodates both clinical and functional limitations (Cooper 2001). The target RPE zone is estimated during a symptom-limited GXT and then produced during individual therapeutic sessions as described in chapter 5. The upper limit of the training zone for pulmonary patients typically is established using a combination of clinical symptoms and physiological indications of impaired alveolar oxygen diffusion or circulatory oxygen transport during dynamic exercise. The target RPE corresponding to the therapeutic training zone is normally set at an exercise intensity just below or at the intensity that causes systemic oxygen insufficiency and provokes intolerable dyspnea. For most pulmonary patients, this symptom-limited training zone is equivalent to an RPE of 4 on both the Borg CR-10 and OMNI scales and corresponds to approximately 40% of patients' symptom-limited maximum effort. For patients with more severe disease (that is, a low FEV_1) the RPE zone should be lowered until dyspnea is tolerable. Patients with mild lung impairment may be able to use a therapeutic RPE training zone of as high as 7 on the CR-10 and OMNI scales. Adjustments in the target RPE zone for pulmonary patients should be made using clinical and physiological information from (a) periodic GXTs, (b) day-to-day training sessions, or (c) quick spot checks of training progress performed using tracking procedures described in chapter 8. When patients experience difficulty in achieving or maintaining the prescribed RPE zone, special coaching, as described in chapter 3, may be helpful. Cooper (2001) developed a comprehensive checklist of the factors that set the upper limits of an RPE training zone for pulmonary exercise rehabilitation.

Pulmonary Rehabilitation Using RPE Zones with Resistance Exercise

Table 9.4 lists target muscle groups and exercises that can be used in pulmonary rehabilitation programs. Each exercise is performed for one or two sets at the muscular endurance RPE zone of 3 on the OMNI-RES. The RPE zone system is used as follows:

- For each exercise, select the desired RPE zone, such as zone 3 for endurance.
- Have the client select, by trial and error, the training weight that produces the target RPE zone.
- Begin the exercise.
- Have the client continue to exercise until reaching an OMNI RPE of 10, when the set is complete.
- Move to the next resistance exercise.

Remember that one exercise repetition includes a complete contraction and relaxation of the target muscle group against the resistance. The OMNI scale should be in full view of the client at all times.

Table 9.4   RPE Zones for Pulmonary Exercise Rehabilitation

Muscle or muscle group	Exercises	Days/week	OMNI RPE zone
Quadriceps	Leg press, leg extension, stand up from chair	2–3	3
Hamstrings	Leg curl*, partial lunge	2–3	3
Triceps surae	Heel raise, seated calf press	2–3	3
Tibialis anterior	Seated ankle dorsiflexion (manual or pulley resistance)	2–3	3
Pectoralis major	Seated chest press*, vertical fly*	2–3	3
Latissimus dorsi	Seated rowing, latissimus pull	2–3	3

(continued)

Table 9.4 (continued)

Muscle or muscle group	Exercises	Days/week	OMNI RPE zone
Deltoids	Seated shoulder press, dumbbell, lateral raise	2–3	3
Trapezius and rhomboids	Reverse fly, shoulder shrug, dumbbell row	2–3	3
Abdominal	Seated crunch with elastic resistance	2–3	3
Biceps	Dumbbell curl	2–3	3
Triceps	Seated dip, triceps press-down	2–3	3

* A seated rather than prone or supine position for these exercises is more easily tolerated by patients with COPD.

Adapted, by permission, from C.B. Cooper, 2001, "Exercise in chronic pulmonary disease: Aerobic exercise prescription," *Medicine and Science in Sports and Exercise* 33(7): S671-S679.

Chest Rating for the Pulmonary Disease Patient

The reference RPE that is used for both clinical assessment and therapeutic prescription in pulmonary rehabilitation can be estimated for either the overall body or the chest. However, the sensitivity of pulmonary exercise testing and exercise rehabilitation is generally better if the patient is asked specifically to rate exertional perceptions arising from the chest. A differentiated RPE-C is directly linked to respiratory function during dynamic exercise and is diagnostically reproducible during repeated exercise tests (Noble and Robertson 1996).

CATEGORY SCALES FOR EXERTIONAL AND CLINICAL PAIN

Sensations of pain that arise during exercise can have clinical, physiological, or psychological causes. The intensity of the pain is influenced by the patient's medical status and fitness level, the type of activity being performed, and the client's pain tolerance and perceptual style. This section discusses three types of pain that are encountered during exercise—cardiovascular (angina), peripheral vascular (claudication), and neuromuscular. Strategies are presented to help both practitioner and client use RPE to manage exercise-induced pain (Noble and Robertson 1996).

Measuring Anginal Pain During Exercise Testing

The intensity of anginal pain during exercise can be measured with the same category scale that is used to determine RPE, usually the Borg CR-10 Scale (Noble and Robertson 1996). The advantage of doing so is that the patient uses a single scale format to estimate the intensity of both anginal symptoms and physical effort, simplifying measurement and making it possible to scale both anginal pain and RPE at the same time points throughout a GXT. If they are given clear scaling instructions before the GXT, patients are able to reliably distinguish between and differentially rate the intensity of anginal pain and perceived exertion.

Rating anginal pain during a GXT has two functions. First, an anginal pain rating determined at a given submaximal exercise intensity can be used to track the patient's progress during exercise rehabilitation. Reduced anginal pain for the same level of work indicates that therapeutic progress is being made. Second, using a predetermined anginal rating as a test termination criterion permits practitioners to discriminate between patients with varying degrees of disease severity and symptom-limited exercise tolerance. Typically, it is recommended that a test of functional exercise tolerance be terminated when the patient gives an *anginal rating* of 7 on the Borg CR-10 Scale. Using this termination criterion, patients with the least advanced disease will have the highest symptom-limited functional aerobic power, whereas patients with the most advanced disease will have the lowest. A uniform test-termination criterion that uses a standard rating of anginal pain facilitates both intra- and interindividual comparison of symptom-limited functional exercise tolerance.

Anginal Pain and Therapeutic Training

Patients with exertional angina pectoris present with a clinical reference point on which a therapeutic exercise prescription can be developed. It is often necessary to define a coronary patient's target training zone according to the perceived intensity of anginal pain (Noble and Robertson 1996). Exercise-induced anginal pain is an ischemic symptom secondary to coronary atherosclerosis. It presents as pain, tightness, or discomfort in the chest, back, jaw, or arms. The sensation is often characterized as radiating from a focal point. Angina pectoris is a complex perceptual experience that includes frank sensation of pain as well as emotional and cognitive responses.

The prescription of a target training intensity for the angina patient should be based on the appearance of pain. As a result of a built-in sensory alarm system that is triggered by the onset of anginal pain, the intensity of exercise therapy can be maintained within a clinically safe and physiologically effective training zone for these patients. The appearance of anginal pain usually coincides with the exercise intensity that surpasses the ischemic threshold. Therefore, exercise should be performed at an intensity that elicits a perceived anginal rating equivalent to the ischemic threshold. Rarely does this target anginal rating exceed a rating of 6 on the Borg CR-10 Scale.

Peripheral Vascular Claudication Pain

Patients with peripheral vascular disease often experience localized intermittent pain in the limbs that are active during exercise due to transient claudication and muscular ischemia. Measurement of tolerable claudication pain can be used to establish the training threshold for these patients' therapeutic exercise rehabilitation. The exercise intensity at which peripheral claudication pain is first detected also serves as a measure of therapeutic progress during rehabilitation. Over time, an increase in the exercise intensity at the claudication threshold indicates therapeutic progress. In this application, the perception of localized muscle pain may be a more salient signal of training progress than RPE (Cook et al. 1997).

NEUROMUSCULAR PAIN

Naturally occurring pain, especially in the limbs and chest, is commonly experienced during high-intensity exercise or when exercise is undertaken for a prolonged period of time. Therefore, practitioners may want to measure clients' pain sensation in conjunction with their RPE to provide a more comprehensive assessment of exercise strain. In general, ratings of pain are moderately correlated with RPE at medium and high exercise intensities. At lower intensities, it may be hard for clients to distinguish between frank pain and more generalized exertional discomfort. In the absence of overt tissue damage, it is not entirely clear what mechanisms are responsible for exercise-induced pain. Mechanical pressure and localized muscular ischemia with consequent production of lactate, hydrogen ions, potassium, and bradykinin may be involved, especially at higher exercise intensities (Cook et al. 1997). Not coincidentally, a number of these biochemical responses are also mediators of peripheral RPE during both aerobic and resistance exercise. However, exercise-induced pain and exertional perception are distinct sensory domains. Therefore, pain and RPE can be measured as separate sensory responses that occur more or less simultaneously during and after exercise.

Pain is typically defined as an unpleasant sensation associated with actual or potential tissue damage. However, clinical cases are reported in which pain is present without apparent tissue damage. When assessing pain during exercise, it is important that the measurement indicate the pain's anatomical location, intensity, pattern of radiation, and time course (Borg 1998). For example, pain can be sharp and short or dull and long. It may appear as a weak, background sensation or an intense, sudden sensation. Clients use a variety of words to describe their exercise-induced pain, including *dull, stinging, pricking, tingling, shooting, numbing,* and *cramping.*

The two category scales most commonly used to measure exercise pain are the Borg CR-10 and Pain Intensity Scales. The CR-10 does not need modification to be used to measure pain sensation. This permits the use of a single scale format to simultaneously assess pain and RPE in patients with musculoskeletal disease who are undergoing prescribed exercise therapy (Borg 1998). However, when using a single category scale to measure two separate perceptions during the same exercise test, clients must have a clear understanding of the differences in the sensory attributes that are to be rated. This information should be provided before exercise begins and restated as needed during the exercise test or conditioning session.

The Pain Intensity Scale (see appendix A) was developed by Cook and colleagues (1997), who recognized that a single measurement tool was needed for both clinical practice and experimental research. This scale combines the numerical format of the Borg CR-10 Scale with verbal descriptors that range from 0 for *no pain at all* to 10 for *extremely intense pain (almost unbearable).* A large dot linked to the verbal descriptor *unbearable pain* appears at the high-intensity end of the response continuum. This scale is a valid and reliable tool for assessing pain during exercise and during the immediate (3 to 6 min) postexercise period for both female and male clients. It should be used as a companion to standard RPE scales when the test objective is to simultaneously measure pain symptoms and exertional perceptions in patients with neuromuscular disease and chronic fatigue syndrome. In these patients, an exercise prescription that simultaneously takes into account pain and exertional symptoms enhances both the physiological efficacy and clinical safety of the program. Instructions for and anchoring procedures to be used with the Pain Intensity Scale appear in appendix B (O'Connor and Cook 2001).

Measurements of naturally occurring muscle pain during exercise have a number of uses, including (a) understanding the factors that influence adoption and maintenance of exercise training, (b) identifying RPE mediators, (c) establishing limits of athletic performance, and (d) developing therapeutic exercise programs for cardiovascular disease and postexercise analgesia (Cook et al. 1997). Exercise-induced pain influences the client's choice of training intensity and decision to continue or discontinue performance (O'Connor and Cook 2001). Athletes often use predetermined pain-rating levels to set their training intensity at the appropriate overload stimulus. In fact, the ability to tolerate naturally occurring muscular pain is considered by some coaches to be a critical factor in determining successful performance in endurance sports (Cook et al. 1997).

SUMMARY OF RATING OF PERCEIVED EXERTION CLINICAL APPLICATIONS

Table 9.5 lists chronic diseases and disabilities amenable to exercise management using RPE. The table provides guidelines for two clinical applications of RPE. First, for each condition, it states whether RPE can be used in combination with physiological and clinical data to assess patient tolerance during a GXT. Second, where available, the table identifies the recommended therapeutic RPE training zone to be targeted for a given disease. These recommendations are summarized from the text of the American College of Sports Medicine's *Exercise Management for Persons with Chronic Diseases and Disabilities* (1997).

Table 9.5   Clinical Applications of RPE in Exercise Testing and Prescription

		RPE therapeutic zone	
Disease	GXT measurement	Borg 6– 20 Scale	OMNI scale conversion
Myocardial infarction	Yes	11–16	4–7
CABG or PTCA	Yes	12–14	5–6
Angina or silent ischemia	Yes	—	—
Pacemaker or ICD	Yes	—	—
Valvular heart disease	Yes	11–14	4–7
Congestive heart failure	Yes	11–16	4–7
Cardiac transplant	Yes	11–16	4–7
Hypertension	Yes	11–13	4–5
Pulmonary disease	Yes	11–13	4–5
Cystic fibrosis	Yes	—	—

Renal failure	Yes	13	5
Diabetes	Yes	—	—
Hyperlipidemia	Yes	11–16	4–7
Obesity	Yes	10–15	3–7
Frailty	Yes	—	—
Anemia	Yes	11–13	4–5
AIDS	Yes	10–14	3–6
Organ transplant	Yes	—	—
Chronic fatigue syndrome	Yes	9–12	2–5
Arthritis	Yes	11–16	4–7
Osteoporosis	Yes	—	—
Stroke or head injury	Yes	—	—
Spinal cord injury	Yes	—	—
Muscular dystrophy	Yes	—	—
Epilepsy	Yes	—	—
Multiple sclerosis	Yes	—	—
Polio or fast polio syndrome	Yes	—	—
Amyotrophic lateral sclerosis	Yes	—	—
Cerebral palsy	Yes	—	—
Parkinson's disease	Yes	—	—

CABG = coronary artery bypass grafting; PTCA = percutaneous transluminal coronary angioplasty; ICD = intracoronary device; — = not available.

From American College of Sports Medicine, 1997, *Exercise management for persons with chronic diseases and disabilities* (Champaign, IL: Human Kinetics).

SUMMARY

Category rating scales for the clinical assessment of cardiac impairment, breathlessness, and pain can be used in combination with the OMNI and Borg perceived exertion scales. Therapeutic exercise prescriptions for coronary, pulmonary, and pain patients should target an RPE training zone that reflects the exertional limitations imposed by both disease severity and functional fitness. The use of RPE indices as adjuncts to physiological and clinical measurements during exercise testing is discussed. Finally, a summary of the clinical applications of RPE assessment is presented for a wide range of disease states.

Appendix A
Category Rating Scales

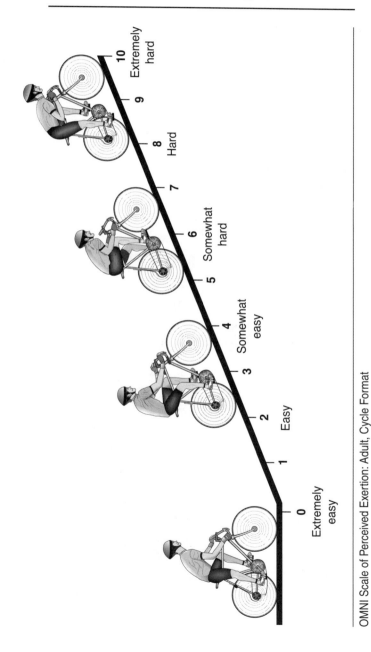

OMNI Scale of Perceived Exertion: Adult, Cycle Format

From *Perceived Exertion for Practitioners: Rating Effort With the OMNI Picture System* by R.J. Robertson. Champaign, IL: Human Kinetics, 2004.

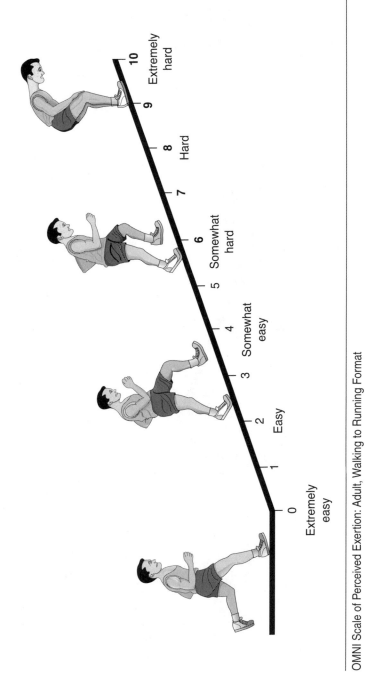

OMNI Scale of Perceived Exertion: Adult, Walking to Running Format

From *Perceived Exertion for Practitioners: Rating Effort With the OMNI Picture System* by R.J. Robertson. Champaign, IL: Human Kinetics, 2004.

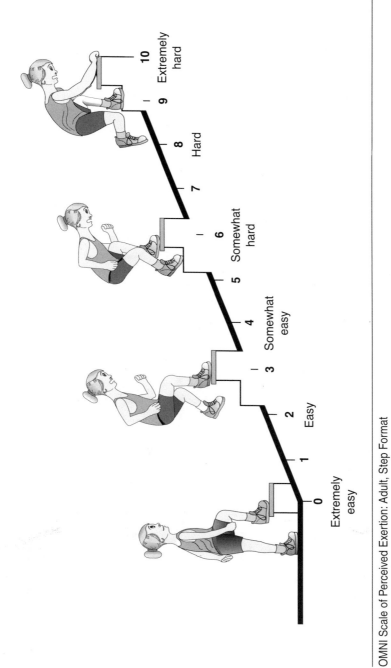

OMNI Scale of Perceived Exertion: Adult, Step Format

From *Perceived Exertion for Practitioners: Rating Effort With the OMNI Picture System* by R.J. Robertson. Champaign, IL: Human Kinetics, 2004.

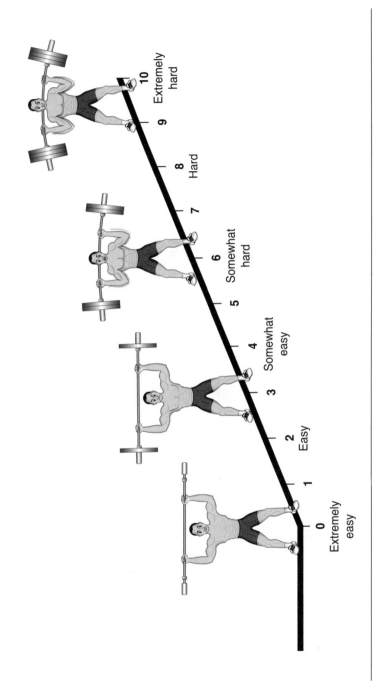

OMNI Resistance Exercise Scale: Adult

From *Perceived Exertion for Practitioners: Rating Effort With the OMNI Picture System* by R.J. Robertson. Champaign, IL: Human Kinetics, 2004.

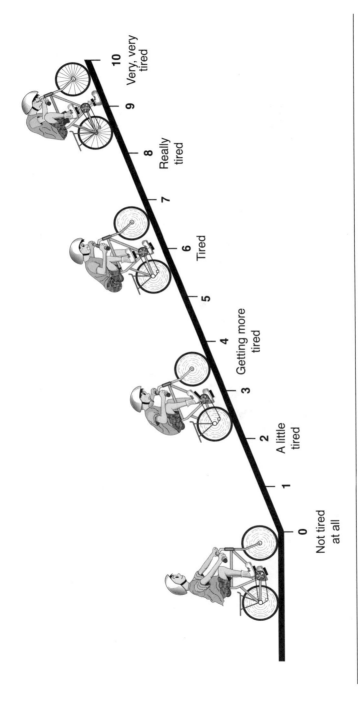

OMNI Scale of Perceived Exertion: Child, Cycle Format

From *Perceived Exertion for Practitioners: Rating Effort With the OMNI Picture System* by R.J. Robertson. Champaign, IL: Human Kinetics, 2004.

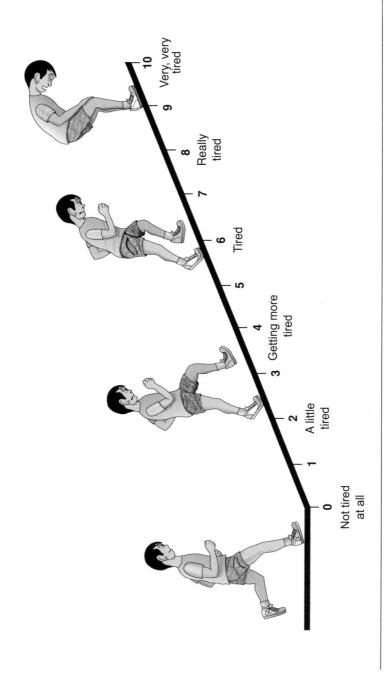

OMNI Scale of Perceived Exertion: Child, Walking to Running Format

From *Perceived Exertion for Practitioners: Rating Effort With the OMNI Picture System* by R.J. Robertson. Champaign, IL: Human Kinetics, 2004.

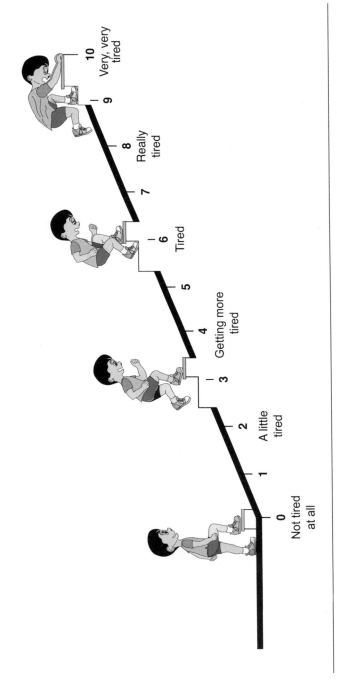

OMNI Scale of Perceived Exertion: Male Child, Step Format

From *Perceived Exertion for Practitioners: Rating Effort With the OMNI Picture System* by R.J. Robertson. Champaign, IL: Human Kinetics, 2004.

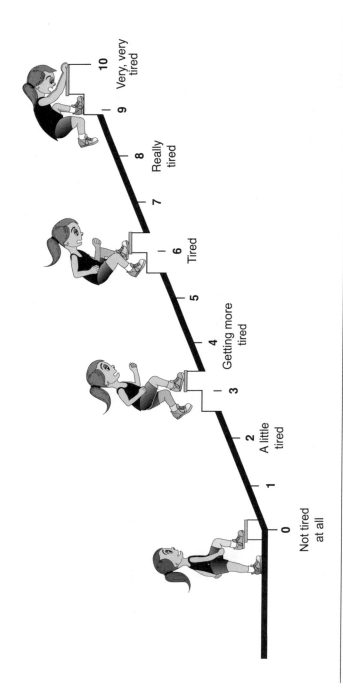

OMNI Scale of Perceived Exertion: Female Child, Step Format

From *Perceived Exertion for Practitioners: Rating Effort With the OMNI Picture System* by R.J. Robertson. Champaign, IL: IL: Human Kinetics, 2004.

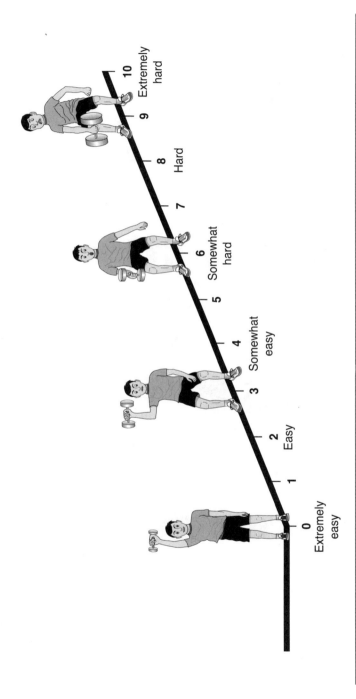

OMNI Resistance Exercise Scale: Male Child

From *Perceived Exertion for Practitioners: Rating Effort With the OMNI Picture System* by R.J. Robertson. Champaign, IL: Human Kinetics, 2004.

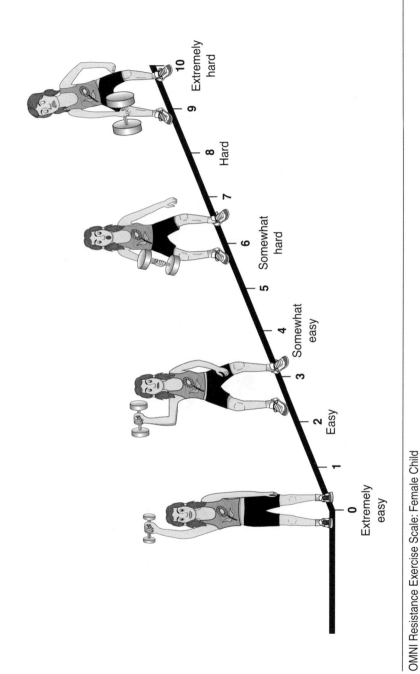

OMNI Resistance Exercise Scale: Female Child

From *Perceived Exertion for Practitioners: Rating Effort With the OMNI Picture System* by R.J. Robertson. Champaign, IL: Human Kinetics, 2004.

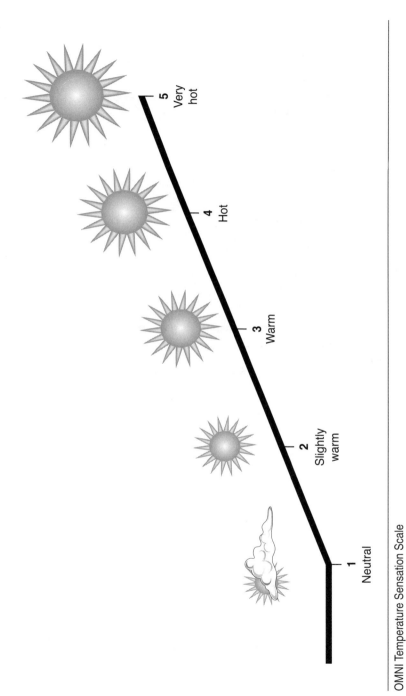

OMNI Temperature Sensation Scale

From *Perceived Exertion for Practitioners: Rating Effort With the OMNI Picture System* by R.J. Robertson. Champaign, IL: Human Kinetics, 2004.

0	Nothing at all	"No P"
0.3		
0.5	Extremely weak	Just noticeable
1	Very weak	
1.5		
2	Weak	Light
2.5		
3	Moderate	
4		
5	Strong	Heavy
6		
7	Very strong	
8		
9		
10	Extremely strong	"Max P"
11		
$\varkappa$		
•	Absolute maximum	Highest possible

Borg CR10 scale
© Gunnar Borg, 1981, 1982, 1998

Borg Category-Ratio Scale (CR-10)

Reprinted, by permission, from G. Borg, 1998, *Perceived exertion and pain scales*. (Champaign, IL: Human Kinetics), 50.

1	
2	Not at all stressful
3	
4	
5	
6	
7	
8	Very, very stressful
9	

University of Pittsburgh Nine-Category Perceived Exertion Scale

1	Very, very easy
2	Very easy
3	Easy
4	Moderate
5	Hard
6	Very hard
7	Very, very hard

Morgan's Seven-Point Category Scale

Adapted, by permission, from W. Morgan, 1985, "Utility of exertional perception with special reference to underwater exercise," *International Journal of Sport Psychology* 32(2): 137-161.

7	Very, very hard
6	Very hard
5	Hard
4	Somewhat hard
3	Fairly light
2	Very light
1	Very, very light

Fleischman's Occupational Effort Index

1	Very, very easy
2	Very easy
3	Easy
4	Just feeling a strain
5	Starting to get hard
6	Getting quite hard
7	Hard
8	Very hard
9	Very, very hard
10	So hard I'm going to stop

Children's Effort Rating Table

Reproduced with permission of authors and publisher from:
Williams, J.G., Eston, R., and Furlong, B. CERT: A perceived exertion scale for young children. *Perceptual and Motor Skills,* 1994, 79, 1451-1458. © Perceptual and Motor Skills 1994.

0	No pain at all
0.5	Very faint pain (just noticeable)
1	Weak pain
2	Mild pain
3	Moderate pain
4	Somewhat strong pain
5	Strong pain
6	
7	Very strong pain
8	
9	
10	Extremely intense pain (almost unbearable)
•	Unbearable pain

Cook's Pain Intensity Scale

Adapted, by permission, from D.B. Cook et al., 1997, "Naturally occurring muscle pain during exercise: Assessment and experimental evidence," *Medicine and Science in Sports and Exercise* 29(8): 999-1012.

Appendix B Rating Scale Instructions

The following are instructions for using the children's OMNI Scale of Perceived Exertion (Robertson et al. 2000a).

Definition

How tired does your body feel during exercise?

Instructions

I would like you to ride on the bicycle for a little while. Every few minutes, it will get harder to pedal the bicycle. Please use the numbers on this picture to tell me how your body feels when you are bicycling. Please look at the person at the bottom of the hill who is just starting to ride a bicycle *(point to the left-hand picture)*. If you feel like this person looks when you are riding, you will be *not tired at all*. You should point to the 0. Now look at the person who is barely able to ride a bicycle to the top of the hill *(point to the right-hand picture)*. If you feel like this person looks when you are riding, you will be *very, very tired*. You should point to the number 10. If you feel like you are somewhere between *not tired at all* (0) and *very, very tired* (10), then point to a number between 0 and 10.

I will ask you to point to the number that tells how your whole body feels, then to the number that tells how your legs feel, and then to the number that tells how your breathing feels. There are no right or wrong numbers. Use both the pictures and the words to help you select the numbers. Use *any* of the numbers to tell how you feel when you are riding the bicycle.

The following are instructions for using the OMNI Resistance Exercise Scale, or OMNI-RES (Robertson et al. 2003).

Definition

Perceived physical exertion is the subjective intensity of effort, strain, discomfort, or fatigue that you feel during exercise.

Instructions

I would like you to use these pictures to describe how your body feels during weightlifting exercise. You are going to perform resistance exercises using your upper and lower body. Please look at the person at the bottom of the scale who is performing a repetition with a light weight. If you feel like this person looks when you are lifting weights, the exertion will be *extremely easy*. When I ask you how you feel, you should respond with the number 0. Now look at the person at the top of the scale who is barely able to perform a repetition with a very heavy weight. If you feel like this person looks when you are lifting weights, the exertion will be *extremely hard*. When I ask you how you feel, you should respond with the number 10. If you feel like your effort is somewhere between *extremely easy* (0) and *extremely hard* (10), then respond with a number between 0 and 10. I will ask you to give a number that describes how your active muscles feel and then a number that describes how your whole body feels. There are no right or wrong numbers. The number you choose may change as you continue to lift weights. Use both the pictures and the words to help you select the numbers. Use any of the numbers to describe how you feel when lifting weights.

The following are instructions for using the Pain Intensity Scale (O'Connor and Cook 2001).

Definition

For this task, *pain* is defined as the intensity of hurt that you feel in your quadriceps muscles only.

Instructions

You are about to undergo a maximal exercise test. The scale before you contains the numbers 0 to 10. Use this scale to describe the feeling of pain in your quadriceps during the exercise test. Don't underestimate or overestimate the degree of pain you feel, just try to estimate it as honestly and objectively as possible. The numbers on the scale represent a range of pain intensity, from *very faint pain* at 0.5 to *extremely intense pain (almost unbearable)* at 10. When you feel no pain in your quadriceps, respond with the number 0. You can also respond with a number greater than 10. If the pain is greater than a 10, respond with the number that represents the pain intensity you feel in comparison to a 10. In other words, if the pain is twice as great as it felt when it was a 10, respond with the number 20. Repeatedly during the test, you will be asked to rate the feelings of pain in your quadriceps. When you rate these pain sensations, be sure to pay attention only to the sensations in your quadriceps and not allow other pains you may be feeling (such as seat discomfort) to influence your rating. It is very important that your rating of pain intensity reflect only the degree of hurt you feel in your quadriceps. Do not use your rating as an expression of fatigue (inability of your quadriceps to produce force) or exertion (how much effort you are putting into performing the exercise).

References

American College of Sports Medicine (ACSM). 1997. *Exercise management for persons with chronic diseases and disabilities.* Champaign, IL: Human Kinetics.

American College of Sports Medicine. 1998. Position stand on exercise and physical activity for older adults. *Medicine and Science in Sports and Exercise* 30, 992-1008.

American College of Sports Medicine. 2000. *Guidelines for exercise testing and prescription.* 6th ed. Baltimore: Lippincott Williams & Wilkins.

American College of Sports Medicine. 2002. Stability balls: an injury risk for older adults. *Health and Fitness Journal* 6, 14-17.

Aquatic Exercise Association. 1995. *Aquatic fitness professional manual.* Nokomis, FL: Aquatic Exercise Association.

Auble, T.E., L. Schwartz, and R.J. Robertson. 1987. Aerobic requirements for moving handweights through various ranges of motion while walking. *Physician and Sportsmedicine* 15, 133-140.

Bar-Or, O. 1977. Age related changes in exercise prescription. In *Physical work and effort,* edited by G. Borg, 255-266. New York: Pergamon Press.

Bonder, B.R., and M.B. Wagner. 2001. *Functional performance in older adults.* 2d ed. Philadelphia: Davis.

Borg, G. 1998. *Borg's perceived exertion and pain scales.* Champaign, IL: Human Kinetics.

Buckworth, J., and R.K. Dishman, eds. 2002. *Exercise psychology.* Champaign, IL: Human Kinetics.

Bushman, B.A., M.G. Flynn, F.F. Andres, L.P. Lambert, M.S. Taylor, and W.A. Braun. 1997. Effect of 4 wk of deep water run training on running performance. *Medicine and Science in Sports and Exercise* 29, 694-699.

Cafarelli, E. 1988. Force sensation in fresh and fatigued human skeletal muscle. *Exercise and Sport Science Reviews* 16, 139-168.

Cook, D.B., P.J. O'Connor, S.A. Eubanks, J.C. Smith, and M. Lee. 1997. Naturally occurring muscle pain during exercise: Assessment and experimental evidence. *Medicine and Science in Sports and Exercise* 29, 999-1012.

Cooper, C.B. 2001. Exercise in chronic pulmonary disease: Aerobic exercise prescription. *Medicine and Science in Sports and Exercise* 33(Suppl. 7), S671-S679.

Covey, M.K., J.L. Larson, S.E. Wirtz, J.K. Berry, N.J. Pogue, C.G. Alex, and M. Patel. 2001. High-intensity inspiratory muscle training in patients with chronic obstructive pulmonary disease and severely reduced function. *Journal of Cardiopulmonary Rehabilitation* 21, 231-240.

Davies, C.T., and A.J. Sargeant. 1979. The effects of atropine and practolol on the perception of exertion during treadmill exercise. *Ergonomics* 22, 1141-1146.

Dishman, R.K. 1994. Prescribing exercise intensity for healthy adults using perceived exertion. *Medicine and Science in Sports and Exercise* 26, 1087-1094.

Doherty, M., P.M. Smith, M.G. Hughes, and D. Collins. 2001. Rating of perceived exertion during high-intensity treadmill running. *Medicine and Science in Sports and Exercise* 33, 1953-1958.

Duncan, G.E., A.D. Mahon, J.A. Gay, and J.J. Sherwood. 1996. Physiological and perceptual responses to graded treadmill and cycle exercise in male children. *Pediatric Exercise Science* 8, 251-258.

Eng, J.J., K.S. Chu, A.S. Dawson, C.M. Kim, and K.E. Hepburn. 2002. Functional walk tests in individuals with stroke: Relation to perceived exertion and myocardial exertion. *Stroke* 33, 756-761.

Eston, R., and D. Connolly. 1996. The use of ratings of perceived exertion for exercise prescription in patients receiving beta-blocker therapy. *Sports Medicine* 21, 176-190.

Feigenbaum, M.S., and M.L. Pollock. 1999. Prescription of resistance training for health and disease. *Medicine and Science in Sports and Exercise* 31, 38-45.

Foster, C., L.I. Hector, R. Welsch, M. Schrager, M.A. Green, and A.C. Snyder. 1995. Effects of specific vs. cross training on running performance. *European Journal of Applied Physiology* 70, 367-372.

Franco, M.J., E.M. Olmstead, A.N.A. Tosteson, D.A. Mahler, T. Lentine, and J. Ward. 1998. Comparison of dyspnea ratings during submaximal constant work exercise with incremental testing. *Medicine and Science in Sports and Exercise* 30, 479-482.

Frangolius, D.D., and E.C. Rhodes. 1995. Maximal and ventilatory threshold responses to treadmill and water immersion running. *Medicine and Science in Sports and Exercise* 27, 1007-1013.

Gamberale, F., A.S. Ljungberg, G. Annwal, and A. Kilbom. 1987. An experimental evaluation of psychophysical criteria for repetitive lifting work. *Applied Ergonomics* 18, 311-321.

Gearhart, R.F., F.L. Goss, K.M. Lagally, J.M. Jakicic, J. Gallagher, and R.J. Robertson. 2001. Standardized scaling procedures for rating perceived exertion during resistance exercise. *Journal of Strength and Conditioning Research* 15, 320-325.

Goss, F.L., R.J. Robertson, S. DaSilva, R. Suminski, J. Kang, and K. Metz. 2003. Ratings of perceived exertion and energy expenditure during light to moderate activity. *Perceptual and Motor Skills* 96, 739-747.

Greenberg, J.S., G.B. Dintiman, and B.M. Oakes. 1998. *Physical fitness and wellness.* 2d ed. Boston: Allyn & Bacon.

Hoffman, J. 2002. *Physiological aspects of sport training and performance.* Champaign, IL: Human Kinetics.

Hogan, J.C., and E.A. Fleischman. 1979. An index of the physical effort required in human task performance. *Journal of Applied Psychology* 64, 197-204.

Jakicic, J.M., B.H. Marcus, K.I. Gallagher, M. Napolitano, and W. Lang. 2003. Effect of exercise duration and intensity on weight loss in overweight, sedentary women: A randomized trial. *JAMA: Journal of the American Medical Association* 290, 1323-1330.

Kreider, R.B., A.C. Fry, and M.L. O'Toole, eds. 1998. *Overtraining in sport.* Champaign, IL: Human Kinetics.

Lagally, K.M., R.J. Robertson, R. Gearhart, K.I. Gallagher, and F.L. Goss. 2002. Ratings of perceived exertion during low and high-intensity resistance exercise in young adults. *Perceptual and Motor Skills* 94, 723-731.

Linderholm, H. 1986. Perceived exertion during exercise in the discrimination between circulatory and pulmonary disorders. In *Perception of exertion in physical work,* edited by G. Borg and D. Ottoson, 199-206. London: Macmillan.

Luxbacher, J., L. Bonci, and K. King. 2002. *Total fitness for women.* Terre Haute, IN: Wish.

Mahler, D.A., and M.B. Horowitz. 1994. Perception of breathlessness during exercise in patients with respiratory disease. *Medicine and Science in Sports and Exercise* 26, 1078-1081.

Mahon, A.D., J.A. Gay, and K.Q. Stolen. 1998. Differentiated ratings of perceived exertion at ventilatory threshold in children and adults. *European Journal of Applied Physiology* 18, 115-120.

Maresh, C.M., M.R. Deschenes, R.L. Seip, L.E. Armstrong, K.L. Robertson, and B.J. Noble. 1993. Perceived exertion during hypobaric hypoxia in low- and moderate altitude natives. *Medicine and Science in Sports and Exercise* 25, 945-951.

Morgan, W.P. 2001. Utility of exertional perception with special reference to underwater exercise. *International Journal of Sport Psychology* 32, 137-161.

Morgan, W.P., D.L. Costill, M.G. Flynn, J.S. Raglin, and P.J. O'Connor. 1988. Mood disturbance following increased training in swimmers. *Medicine and Science in Sports and Exercise* 20, 408-414.

Morgan, W.P., and M.L. Pollock. 1977. Psychological characteristics of elite runners. *Annals of the New York Academy of Science* 301, 382-403.

Moyna, N.M., R.J. Robertson, C.L. Meckes, J.A. Peoples, N.B. Millich, and P.D. Thompson. 2001. Intermodal comparison of energy expenditure at exercise intensities corresponding to the perceptual preference range. *Medicine and Science in Sports and Exercise* 33, 1404-1410.

Noble, B.J., and R.J. Robertson. 1996. *Perceived exertion.* Champaign, IL: Human Kinetics.

O'Connor, P.J., and D.B. Cook. 2001. Moderate intensity muscle pain can be produced and sustained during cycle ergometry. *Medicine and Science in Sports and Exercise* 33, 1046-1051.

O'Connor, P.J., M.S. Poudevigne, and J.D. Pasley. 2002. Perceived exertion responses to unaccustomed elbow flexor eccentric actions in women and men. *Medicine and Science in Sports and Exercise* 34, 862-868.

Pandolf, K.B. 2001. Rated perceived exertion during exercise in the heat, cold or at high altitude. *International Journal of Sport Psychology* 32, 162-176.

Pandolf, K.B., and B.J. Noble. 1973. The effect of pedaling speed and resistance changes on perceived exertion for equivalent power outputs on the bicycle ergometer. *Medicine and Science in Sports and Exercise* 5, 132-136.

Pfeiffer, K.A., J.M. Pivarnik, C.J. Womack, M.J. Reeves, and R.M. Malina. 2002. Reliability and validity of the Borg and OMNI RPE Scales in adolescent girls. *Medicine and Science in Sports and Exercise* 34, 2057-2061.

Pierce, K., R. Rozenek, and M.H. Stone. 1993. Effects of high volume weight training on lactate, heart rate and perceived exertion. *Journal of Strength and Conditioning Research* 7, 211-215.

Pincivero, D.M., A.J. Coelho, R.M. Campy, Y. Salfetnikov, and A. Bright. 2001. The effects of voluntary contraction intensity and gender on perceived exertion during isokinetic quadriceps exercise. *European Journal of Applied Physiology* 84, 221-226.

Robertson, R.J. 2001a. Development of the perceived exertion knowledge base: An interdisciplinary process. *International Journal of Sport Psychology* 32, 189-196.

Robertson, R.J. 2001b. Exercise testing and prescription using RPE as a criterion variable. *International Journal of Sport Psychology* 32, 177-188.

Robertson, R.J., C.J. Caspersen, T.G. Allison, G.S. Skrinar, R.A. Abbott, and K.F. Metz. 1982. Differentiated perceptions of exertion and energy cost of young women while carrying loads. *European Journal of Applied Physiology* 49, 69-78.

Robertson, R.J., F.L. Goss, T.E. Auble, R. Spina, D. Cassinelli, E. Glickman, R. Galbreath, and K.F. Metz. 1990a. Cross-modal exercise prescription at absolute and relative oxygen uptake using perceived exertion. *Medicine and Science in Sports and Exercise* 22, 653-659.

Robertson, R.J., F.L. Goss, J.A. Bell, C.R. Dixon, K.I. Gallagher, K.M. Lagally, J.M. Timmer, K.L. Abt, J.D. Gallagher, and T. Thompkins. 2002. Self-regulated cycling using the children's OMNI Scale of Perceived Exertion. *Medicine and Science in Sports and Exercise* 34, 1168-1175.

Robertson, R.J., F.L. Goss, N. Boer, J.D. Gallagher, T. Thompkins, K. Bufalino, G. Balasekaran, C. Meckes, J. Pintar, and A. Williams. 2001. OMNI Scale perceived exertion at ventilatory breakpoint in children: Response normalized. *Medicine and Science in Sports and Exercise* 33, 1946-1952.

Robertson, R.J., F.L. Goss, N.F. Boer, J.A. Peoples, A.J. Foreman, I.M. Dabayebeh, N.B. Millich, G. Balasekaran, S.E. Riechman, J.D. Gallagher, and T. Thompkins. 2000a. Children's OMNI Scale of perceived exertion: Mixed gender and race validation. *Medicine and Science in Sports and Exercise* 32, 452-458.

Robertson, R.J., F.L. Goss, M. Dupain, J. Rutkowski, C. Brennan, J. Dube, J. Andreacci, and D. Jenkinson. 2004. Validation of the adult OMNI scale of perceived exertion for cycle ergometer exercise. *Medicine and Science in Sports and Exercise* 36, 102-108.

Robertson, R.J., F.L. Goss, T. Michael, N. Moyna, P. Gordon, P. Visich, J. Kang, T. Angelopoulos, S. DaSilva, and K. Metz. 1995. Metabolic and perceptual responses during arm and leg ergometry in water and air. *Medicine and Science in Sports and Exercise* 27, 760-764.

Robertson, R.J., F.L. Goss, J. Rutkowski, B. Lenz, C. Dixon, J. Timmer, K. Frazee, J. Dube, and J. Andreacci. 2003. Concurrent validation of the OMNI perceived exertion scale for resistance exercise. *Medicine and Science in Sports and Exercise* 35, 333-341.

Robertson, R., K. Metz, F. Goss, J. Kang, C. Sprowls, N. Moyna, S. DaSilva, M. O'Connor, R. Suminski, T. Michael, P. Gordon, and P. Visich. 1994. Validation of a run test to predict maximal oxygen uptake in young men and women using perceived exertion and heart rate as reference variables. *Medicine and Science in Sports and Exercise* 26(Suppl), S166.

Robertson, R.J., N.M. Moyna, K.L. Sward, N.B. Millich, F.L. Goss, and P.D. Thompson. 2000b. Gender comparison of RPE at absolute and relative physiological criteria. *Medicine and Science in Sports and Exercise* 32, 2120-2129.

Robertson, R.J., P.A. Nixon, C.J. Caspersen, K.F. Metz, R.A. Abbott, and F.L. Goss. 1992. Abatement of exertional perceptions following high intensity exercise: Physiological mediators. *Medicine and Science in Sports and Exercise* 24, 346-353.

Robertson, R.J., and B.J. Noble. 1997. Perception of physical exertion: Methods, mediators and applications. *Exercise and Sport Science Reviews* 25, 407-452.

Robertson, R.J., R.T. Stanko, F.L. Goss, R.J. Spina, J.J. Reilly, and K.D. Greenawalt. 1990b. Blood glucose extraction as a mediator of perceived exertion during prolonged exercise. *European Journal of Applied Physiology* 61, 100-105.

Shephard, R.J., T. Kavanagh, D.J. Mertens, and M. Yacoub. 1996. The place of perceived exertion ratings in exercise prescription for cardiac transplant patients before and after training. *British Journal of Sports Medicine* 30, 116-121.

Snyder, A.C., A.E. Jeukendrup, M.K.C. Hesselink, H. Kuipers, and C. Foster. 1993. A physiological/psychological indicator of over reaching during intensive training. *International Journal of Sports Medicine* 14, 29-32.

Svedenhag, J., and J. Seger. 1992. Running on land and in water: Comparative exercise physiology. *Medicine and Science in Sports and Exercise* 24, 1155-1160.

Tikuuisis, P., T.M. Mclellan, and G. Selkirk. 2002. Perceptual versus physiological heat strain during exercise-heat stress. *Medicine and Science in Sports and Exercise* 34, 1454-1461.

U.S. Department of Health and Human Services. 1996. *Physical activity and health: A report of the Surgeon General.* Atlanta, GA: Department of Health and Human Services, Centers for Disease Control and Prevention, National Center for Chronic Disease Prevention and Health Promotion.

Utter, A.C., R.J. Robertson, D. Nieman, and J. Kang. 2002. Children's OMNI scale of perceived exertion: Walking/running evaluation. *Medicine and Science in Sports and Exercise* 34, 139-144.

Van Den Burg, M., and R. Ceci. 1986. A comparison of a psychophysical estimation and a production method in a laboratory and a field condition. In *Perception of exertion in physical work,* edited by G. Borg and D. Ottoson, 35-46. London: Macmillan.

Wathen, D., and F. Roll. 1994. Training methods and modes. In *Essentials of Strength Training and Conditioning,* edited by T. Baechle, 404-415. Champaign, IL: Human Kinetics.

Whaley, M.H., P.H. Brubaker, L.A. Kaminsky, and C.R. Miller. 1997. Validity of rating of perceived exertion during graded exercise testing in apparently healthy adults and cardiac patients. *Journal of Cardiopulmonary Rehabilitation* 17, 261-267.

Wilder, R.P., and D.K. Brennan. 1993. Physiological responses to deep water running in athletes. *Sports Medicine* 16, 374-380.

Index

Note: The italicized *f* and *t* following page numbers refer to figures and tables, respectively.

A

activities of daily living 80, 82
aerobic fitness, assessing
  multilevel cycle test 37-38, 38*t*, 39, 39*f*
  multilevel RPE cycle test 39-40
  RPE run test 41-42, 42*f*, 43
  RPE walk test for clinical assessment 44
  run test nomogram 43, 43*f*, 44
  single-level cycle test 40, 40*t*
  single-level cycle test score, interpreting 41
  submaximal RPE cycle tests 37
age and measuring RPE during exercise testing 50
ambient setting and measuring RPE during exercise testing 51
anaerobic threshold zone 66, 67*f*
anchoring procedures 17-18
anchor points 17
aquatic zones 87-88, 88*t*, 89

B

Borg, Gunnar 2, 22
Borg Category-Ratio Scale (CR-10 scale) 22, 23, 152
Borg 15-category scale 2, 2*f*, 14, 22
Borg 6-20 Scale 23, 24-25, 26
Borg's range model
  anchoring procedures 4
  for category scales 4, 4*f*, 5
  measurement assumptions of 4

C

cardiac medications, effects of 128-129
cardiac patient, exercise therapy for
  exercise prescription, objective of 127, 127*t*, 128*t*
  inpatient and outpatient programs 126
  ischemic threshold and perceived exertion 126
  physiological and clinical assumptions 126
cardiac transplant patients and RPE 129-130

category rating scales
  Borg's perceived exertion and pain scales 22
  history of 22
  individual scales 141-154
  other perceived exertion scales 22
  validating of 22
category RPE scale 8
Center for Exercise and Health-Fitness Research (CEHFR) 39
children and youth, RPE training zones 70
Children's Effort Rating Table (CERT) 22, 153
chronic obstructive pulmonary disease (COPD), dyspnea index for 130-132
client calibration 62
clinical applications of RPE, summary 138, 138*t*-139*t*
club and clinic, health-fitness goals in
  effective program, goals for 64
  helping clients meet goals 64
  promoting adherence 64-65
combination training 76, 76*f*, 77, 77*t*
competition, RPE training zones for
  case study 113
  core rating of 117-118, 118*t*
  core training 114
  overload training principle 114
  overtraining syndrome, RPE indices of 119-121
  quantity and quality overload training 114-117
  resistance exercise 118
  sport-specific training 114
  tracking training progress 122-123
Cook's pain intensity scale 154
cross-training
  RPE zone system for 69-70
  target RPE zones for 57-58

D

daily activity zone 82
differentiated RPE 8
dominant RPE
  differentiated 30-31

reinforcement of OMNI scaling procedures 18
reinforcing appropriate RPE estimates 19
under-and-overestimated RPE responses, correcting 19
rehabilitation patient, exercise testing and training. *See* exercise testing and training for rehabilitation patient
research subject, solutions for 30
resistance exercise
RPE measurement during 46
RPE training zones for competition 118
for wheelchair client 74, 75, 75*t*
respiratory-metabolic mediators 5
RPE as test guide
guiding test progression 35, 37
measurement, functions of 34
monitoring RPE for increase 35, 37*t*
warning zones 34-35, 35*f*, 36*f*
RPE (rating of perceived exertion)
climbs for exercise duration 69
estimation procedure 54-55, 55*f*, 56
preference zone 84-85, 86*t*, 87*t*
walk test for clinical assessment 44
RPE run test
description of 41
estimating aerobic fitness 42, 42*f*
test procedures 41-42
test results, calculating 42, 43
when to administer 43
RPE scale, anchoring
exercise anchoring procedures 17-18
rating anchor points 17
setting for each client 17
RPE training zone
anaerobic threshold zone 66, 67*f*
defined 8
description of 65
RPE climbs 69
RPE conversion 65, 66*f*
sliding RPE zones 66, 68*f*, 69
RPE training zones for competition
case study 113
core RPE training zones 117-118
core training and sport-specific training 114
overload training principle 114
overtraining syndrome 119-121
quantity and quality overload training 114-117
resistance exercise 118
tracking training progress 122-123
RPE zones for special clients and conditions
case study 95

children's exercise needs 96
environmental influences on 109-110
occupational applications of 110-112
women, older clients, and children 96-108
RPE zone system
activities of daily living 80, 82
aquatic zones 87-88, 88*t*
children and youth 70
combination training 76, 76*f*, 77, 77*t*
cross-training 69-70
energy-efficient RPE zones 85, 86*t*-87*t*, 87
flexibility 92-93
group cycling program 78*t*
home exerciser 92
indoor group cycling 77-78
marathon training 79
mind-body programs 84
preference zone 84-85, 86*t*-87*t*, 87
sliding zone system for resistance training 71, 72*f*, 73-74, 74*t*
spinning program 79*t*
stability ball exercises 82, 83*t*
stretch bands and cords 83-84
training improvements and daily adjustments 70
water running 89, 90*t*
12-week marathon training programs 80*t*, 81*t*
weight loss 89, 91, 91*t*
RPO score 33, 40
run test nomogram
building nomogram 44
nomogram, description of 43
nomogram to predict $\dot{V}O_2$max 43, 43*f*

S

scale anchoring procedures
after test 27
Borg 6-20 RPE scale 26
exercise anchoring protocol 26
high scale anchor 27
low scale anchor 27
scale-anchoring graded exercise test (GXT) 26
scale anchors
exercise anchoring 26
exercise and memory anchoring 26
memory anchoring 26
scaling difficulties
identifying clients 28, 29*f*
rating difficulties *vs.* normal RPE response fluctuations 28, 29
scaling skills
information to present 27-28

About the Author

Robert J. Robertson, PhD, is professor of exercise physiology at the University of Pittsburgh and codirector of the Center for Exercise and Health-Fitness Research. He is responsible for the development, validation, and application of the OMNI Picture System of perceived exertion assessment. Since earning his PhD in exercise physiology from the University of Pittsburgh in 1973, Dr. Robertson has had extensive teaching, research, and writing experience in the field of perceived exertion. The results of his research have been published in refereed journals and presented at national and international conferences.

Dr. Robertson is coauthor of *Perceived Exertion* (Human Kinetics 1996) and associate editor for the psychobiology section of *Medicine and Science in Sports and Exercise.* He is a program director and fellow of the American College of Sports Medicine (ACSM).

You'll find other outstanding fitness resources at

www.HumanKinetics.com

In the U.S. call

1-800-747-4457

Australia................................ 08 8277 1555
Canada 1-800-465-7301
Europe......................+44 (0) 113 255 5665
New Zealand................... 0064 9 448 1207

HUMAN KINETICS
The Information Leader in Physical Activity
P.O. Box 5076 • Champaign, IL 61825-5076 USA

LIBRARY, UNIVERSITY OF CHESTER